SUPERMARINE

ATTACKER

SUPERMARINE

ATTACKER

THE ROYAL NAVY'S FIRST
OPERATIONAL JET FIGHTER

RICHARD A FRANKS

DALRYMPLE
& VERDUN◆
PUBLISHING

Attacker
The Royal Navy's First Operational
Jet Fighter
by Richard A Franks

ISBN (10) 1-905414-05-6
ISBN (13) 978-1-905414-05-5

First published in 2007 by
Dalrymple & Verdun Publishing
33 Adelaide Street, Stamford,
Lincolnshire, PE9 2EN
Tel: 0845 838 1940
mail@dvpublishing.co.uk
www.dvpublishing.co.uk

© Concept and design
Dalrymple & Verdun and
Stephen Thompson Associates
© Richard A Franks 2007
© Richard J Caruana colour profiles

Editor and commissioning editor
Martin Derry

Printed in England by
Ian Allan Printing Ltd
Riverdene Business Park
Molesey Road
Hersham, Surrey, KT12 4RG

Acknowledgments
A word of thanks once again to Barry
Jones for his help with information and
images. Also my thanks go to Richard
J Caruana for his excellent colour profiles.

Publishers acknowledgments
The publishers would like to thank Chris
Salter for his invaluable assistance and
the use of his personal records. Special
thanks also to Jan Keohane and
Catherine Rounsfell of the FAA Museum
archive for their help with research and
photographs. Tony Buttler, Ian Gazeley,
Brian Lowe, Mike Smith, Andy Thomas,
and Brandon White for their generous
and selfless contribution of many
previously unpublished images.

CONTENTS

Half title page: *HMS* Eagle *at Devonport in late May 1954 on her return from a tour of the Mediterranean. 800 and 803 Squadron Attackers stand in the foreground with four 806 Squadron Hawker Sea Hawks to their rear. Grumman Avengers and Douglas Skyraiders stand prominently at the very stern of the flight deck. When returning to home port it was usual for aircraft to be flown off to land bases before entering port, but it was not so on this occasion!* Brian Lowe

Title page: *A photograph of HMS* Eagle *at speed with Attackers on deck accompanied by Douglas Skyraider AEW.1s and Fairey Fireflies, taken between March 1952 and mid-1954. The three-funnelled vessel astern of* Eagle *is one of three surviving fast minelayers (from a total of six) to have survived the Second World War. Therefore the vessel seen here is either* Manxman, Apollo *or* Ariadne. *They were extremely fast, wartime propaganda crediting them with speeds of 42 knots or more, whereas in fact their actual speeds were in the region of 39½ knots, still very fast. Because of this they were used to supply Malta with priority materials during the siege by Axis forces, often running without escort. These vessels were scrapped between 1962 and 1971.* FAA Museum

Left: *An unidentified Attacker is marshalled toward* Eagle's *port catapult with a Fairey Firefly in the foreground. Hovering close by is a Piasecki HUP-2 of the US Navy.* via Ian Gazeley

FOREWORD

Commander Bertie Vigrass
OBE, VRD, RNR(Rtd).

The Attacker was the first operational jet fighter aircraft in service with the Royal Navy. However, it had a relatively short operational life in first-line squadron service – only six years from 1951 to a dramatic end on 10th March 1957 when the 'Sandys' Axe' fell.

Commencing in May 1955, three RNVR Squadrons were equipped with the Attacker: No.1831 Squadron (Southern Air Division), No.1832 (Northern Air Division) and No.1833 (Midland Air Division). I was at that time the Commanding Officer of the Midland Air Division. We were very pleased to be re-equipped with these aircraft as it brought us into the jet age. However it is debatable whether, from an operational flying point of view, the Attackers were an improvement on the Sea Fury which they had replaced.

The pure fighter version of the Attacker was not a great success. Its rate of climb and manoeuvrability at high altitude was disappointing. The fighter-bomber version (FB.2) which the three RNVR Squadrons received were much more suitable for our purposes. These aircraft were fitted with bomb racks, a 250 gallon belly tank and rocket launchers. The design of the aircraft with its tail wheel undercarriage led to a number of problems. The aircraft was certainly more difficult to deck-land than the contemporary nose-wheel naval jet aircraft. It was fortunate that HMS *Eagle* had been installed with the newly invented gyro-stabilised mirror landing sight. This must have reduced the risk of Attacker deck-landing accidents.

We in the Midland Air Division had our own problems related to the tail-wheel configuration of the aircraft. The Air Division consisted of two squadrons – No.1833, the fighter squadron and No.1844 – the torpedo/bomber/reconnaissance squadron. Our base was at RNAS Bramcote in the midlands which had a grass airfield. The airfield had been perfectly suitable for the Air Division's aircraft over the years – the Supermarine Seafire, the Hawker Sea Fury, the Fairey Firefly and the Grumman Avenger, and even the Attacker could be flown safely from the airfield. However, because of the tail-wheel design, the Nene engine blasted hot gasses on to the ground immediately astern of the aircraft. This of course caused permanent damage to the runways and perimeter track.

Our solution was to transfer No.1833 Squadron to RAF Honiley, about 15 miles away. No.1844 Squadron and the headquarters of the Air Division remained at Bramcote. This was far from an ideal situation, but it had the advantage that No.1833 Squadron was able to train with No.605 County of Warwick RAF Auxiliary Squadron which was also based at Honiley.

However, this all came to an abrupt end in March 1957 as a result of the 'Sandy's Axe' and the severe defence cuts. All flying in the RNVR squadrons and the RAF Auxiliary squadrons ceased. This brought the first-line operational life of the Attacker to an end.

Lieutenant Commander D T Chute
DSC, VRD**, RNR(Rtd).

In July 1955, 1833 Squadron RNVR commenced a conversion course to introduce the Supermarine Attacker FB.2 with Rolls-Royce Nene engine in exchange for the Hawker Sea Fury. The first jet experience was given on the De Havilland Sea Vampire T.22 and gave about 25 hours on Vampires and 10 on Attackers. I had 182 hours on the Attacker which I am sure was exceeded by quite a few RN pilots.

Because the Attacker had a tail wheel and a tail down attitude on the ground as opposed to the tricycle undercarriage of later jets such as the Hawker Sea Hawk, the squadron had to move from the grass field with Sommerfeld Tracking at RNAS Bramcote to lodge at RAF Honiley with a Royal Auxiliary Air Force squadron flying single seat Vampires.

The squadron pilots were pleased to progress to jet aircraft and first impressions of the Attacker were that it gave a smooth flight, having none of the torque problems associated with propeller driven aircraft, in particular that produced by the Centaurus 18 of the Sea Fury. Unfortunately the endurance of the new aircraft was not good even when equipped with a large long range fuel tank which fitted closely around the underside of the fuselage. With the tank fitted most flights were of about one hour duration, the longest being about one hour and twenty minutes. The use of a long range tank (LRT) had to be carefully managed and was the cause of the one fatality in the squadron when a pilot failed to switch back to main tanks on approaching the Honiley circuit. His engine cut from lack of fuel and the aircraft crashed in a wood alongside the airfield.

On one occasion in the middle of a flight, the cockpit canopy flew off, but slow flying at altitude showed no untoward handling characteristics and a noisy and cold flight was completed with a safe landing. No report was received of the discovery of the canopy.

As an aerobatic aircraft the Attacker was easy to handle especially when the LRT was not fitted but the endurance was considerably limited.

Bertie Vigrass at RNAS Bramcote in the spring of 1955, with Sea Fury FB.11 WZ656 158/BR. This aircraft, the last FB.11 to be built, had been delivered, new, to 1833 Squadron on 6th October 1954, finally leaving the unit on 22nd June 1955. Bertie Vigrass

There is a newsreel film of an RN squadron Attacker giving a display at an air day where the pilot approached the runway at a very low level, rolled the aircraft into inverted flight and continued down the runway, still at a very low level. One got the impression that the pilot had to lift the aircraft slightly to give clearance for the wing in the roll!

A factor which ex-piston-engine pilots had to realise was that the response to an increase in throttle power was not as immediate nor as lively as they had been used to. This slower response led to a very unfortunate accident involving an Attacker on finals approaching the runway at Honiley and needing to increase height urgently. The aircraft did not respond immediately and it crashed through the boundary hedge. The pilot was unhurt and the aircraft only slightly damaged, but a young lad had been standing on the hedge in line with the runway, 'goofing' at the airfield activities and he was hit and killed.

This lack of immediate response became apparent when the squadron started Mirror Airfield Dummy Deck Landings in preparation for embarking in an aircraft carrier for annual training. As it happened the carrier was required for Anthony Eden's Suez crisis and the squadron never got to sea.

Instead of flight-deck experience the squadron did its last annual training at RNAS Ford where they were able to experience the Attacker as a weapon platform. It so happened that most of the air-towed target banners were lost and there was no real assessment of the air-to-air firing capabilities of the pilots. The firing of air-to-ground rockets was, however, quite successful.

Captain E M 'Winkle' Brown RN was a test pilot of very great experience who first deck landed the Attacker. His opinion was that the first operational naval jet fighter was never a brilliant aircraft but did perform reasonably well in squadron service.

RNVR flying came to an end in early 1957 when all reserve flying was discontinued.

THE ATTACKER IN CONTEXT

Supermarine's Attacker has never faired well as far as published material is concerned. This may reflect service opinion that the Attacker didn't prove to be a great success with the Royal Navy and today, is regarded in retrospect, even by those who know of its existence, as an also-ran (or flew)! Certainly when examining this aircraft it is soon realised that it only ever operated from one aircraft carrier; HMS *Eagle*, (excluding deck landing trials and practise conducted on other vessels), and then only for a relatively short period of time. When exploring *Eagle*'s history the many differing aircraft types operated from her deck during 20 years of service are frequently commented on, piston-engine aircraft included, yet the Attacker is hardly referred to – if at all, despite being the Royal Navy's first operational jet.

Viewed retrospectively the Attacker never had the charisma of other jet fighters, even by early 1950s' standards. By then swept-wing designs were to be found in service within the USA and USSR, with other swept-wing designs appearing in prototype form in the UK, France and elsewhere. Against these the Attacker already looked dated, as indeed it did even if compared to other contemporary 'straight-wing' fighter designs, for example Gloster's Meteor or Dassault's Ouragan, (Hurricane). The most obvious difference between the Attacker and other jet designs was the inclusion of a tail wheel undercarriage, synonymous with the piston-engine era rather than the jet era. Adding to its 'image' problem and imported directly

from Supermarine's piston-engine Spiteful and Seafang, from which the Attacker was partly derived, was its wing-mounted armament. Most British jet fighters carried their guns in the fuselage with the exception of the later Meteor night fighters and the Javelin. As recounted later by Richard, wing mounted guns at least prevented problems of gas ingestion into engine air intakes when being fired. However it has been stated that when the Attacker was involved in mock combat, high levels of stress could be generated causing the wings to flex and distort, in turn pushing the guns out of alignment with obvious consequences for any fighter relying on gun armament alone. The Attacker in the air did look rather sleek initially depending on angle of sight, until that is, it acquired a huge ventral fuel tank, very soon required to overcome a lack of range and endurance when flying on internal fuel alone. This was a problem encountered time and again in British fighter designs, for the most part resolved in the jet era by the inclusion of streamlined under wing or under fuselage drop tanks. The Attacker's external tank carried 250 gallons of fuel in addition to the 293 gallons carried internally, which sounds significant, although it is worth noting in general terms that 70 gallons equated to approximately 10 minutes flying time at altitude.

The Attacker was important though; because it played its part in allowing the Royal Navy to acquire valuable information and knowledge whilst operating jet aircraft at sea, not just for a few days, but for

The first prototype, TS409, at a fairly early stage in its life, seen here with the original small tail, natural metal finish and type 'C' roundels on the wings and 'C1' roundels on the fuselage. Of note are the national markings on the fin, which was not applied to production machines. FAA Museum

extended periods. In operating the Attacker the RN confirmed what it probably already knew; that modifications would be required to their aircraft carriers in order to operate later jets successfully, especially as the aircraft themselves became larger, faster and heavier. In due course the Royal Navy would be the first of the world's navies to introduce vital improvements for landing jets more safely on aircraft carriers at sea. Principally these were the mirror landing aid and angled flight deck, (not forgetting, of course, the steam catapult which was designed to launch ever heavier, faster aircraft).

Jets themselves also required new operating techniques. Because of their high rate of fuel consumption and because less fuel was burned at high altitude, jets would ideally remain above 20,000 feet when returning to their aircraft carrier invariably short of fuel, an altitude from which, in certain weather conditions a pilot may not even be able to see his ship at all! They would then descend rapidly to land on deck as soon as possible with just a few minutes fuel remaining – precious little fuel with which to loiter if, for whatever reason, the pilot was forced 'to go round again'. This was the opposite of piston-engine aircraft which would descend to almost sea level before 'queuing' prior to making their deck recovery. Piston-engine aircraft were of course very much the norm when the Attacker first went to sea. They were relatively economical with fuel and, importantly, their throttle response time was rapid. Jet engines of the period on the other hand were slow to react to throttle movements making descent and landing all the more difficult to judge, and although fitted with ejection seats should things go wrong, early seats required thousands of feet of altitude to be operated safely. The pilot, having ejected would have to manually release himself from the seat and then deploy his parachute, so if when landing aboard matters got out of hand, the ejection seat was of little use. In short, for many reasons the use of jet aircraft from a moving deck in marginal weather could be very dangerous.

For the pilot, having reached the threshold of an aircraft carrier's stern, there remained the urgent need for his arrester hook to engage an arrester wire. If accomplished all well and good, but for those Attacker pilots who missed the wire, a sinister problem arose. Normally if a piston-engine aircraft 'bolted' it would hopefully engage the crash barrier rigged across the ships flight deck, thus preventing the runaway crashing into aircraft parked toward the ships bow, this being before the advent of the angled flight deck which would allow a bolter to open the throttle and accelerate away along an unobstructed deck. As jet engine throttle response time improved, the angled deck would prove its worth time and again. For the piston-engine aircraft, engaging the barrier would cause extensive engine and airframe damage, the barrier being constructed of horizontal steel cables but, with a big engine and propeller in front to absorb

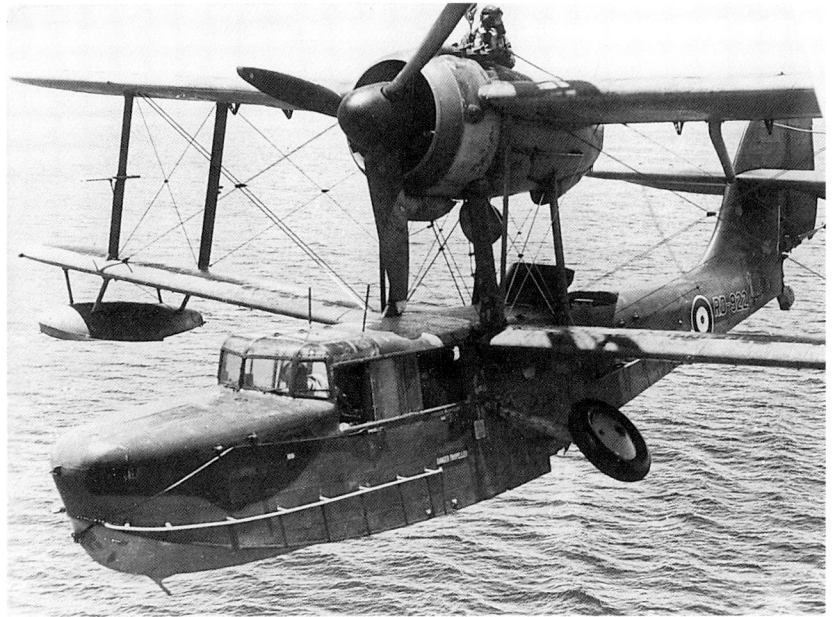

the impact, the pilot would likely survive. The Attacker pilot had no such protection so a nylon crash barrier was developed. This allowed the Attacker's nose and cockpit to pass through the barrier, engaging the wings and causing extensive damage to the airframe, as some of the photographs in this book will testify. However it made such incidents more survivable. Given these hazards it is always worth remembering that although this book is about an aeroplane it is more importantly about the men who flew them and their skill and courage in ushering in a new age of naval aviation in which many pilots were to lose their lives.

As stated, Supermarine's Attacker would not remain with the Royal Navy's regular squadrons for long, largely being replaced by Hawker's Sea Hawk. The Attacker's service life would continue for a while longer with RNVR units, civilian run units and the Royal Pakistan Air Force until early 1956. As for Supermarine, they would produce further jet fighter designs, two of which would be developed for service use: the Scimitar for the Royal Navy and the Swift for the RAF, the latter especially with mixed results.

In retrospect the Attacker has been described as an anachronism which failed to demonstrate the tremendous technical strides and achievements that had occurred in the British aviation industry during the 1940s. Perhaps the Attacker is better described as having been an immature example of a rapidly evolving technology.

Martin Derry
Stamford, February 2007

The Supermarine Sea Otter graphically illustrates the stark differences in technologies in use at the dawn of the jet age. Almost unbelievably Sea Otter ASR.II RD-922, the last of its type to be built, was accepted by the Royal Navy on 1st August 1946, whilst Attacker prototype TS409 (opposite), had made its first flight five days earlier on 27th July 1946. Brian Lowe

Chapter 1: **FROM PISTON TO JET**

A photograph taken in 1946 of a Supermarine Seafang F.31 – most probably VG471. Although its Spitfire heritage is still very evident in many respects, the shape of the wing informs us that it is no longer a Spitfire. Neither the Seafang nor its stablemate the Spiteful would enter service, but the wing, suitably modified, would be incorporated into the jet-powered Attacker.
via Tony Buttler

Before examining the development and service career of the Attacker, two earlier Supermarine aircraft, their Spiteful and Seafang, should be considered. It is often stated that the wings of the Spiteful were incorporated into the Attacker, and to a degree this is correct, with the modified Spiteful wing being used on the first three prototype Attackers. It was however the Seafang's wing, suitably modified, that was actually used in production Attackers. The Spiteful originated from Supermarine's chief designer Joseph Smith, whose replacement Spitfire design to meet Specification F.1/43 resulted in the Spiteful in 1944. Gone were the elliptical wings of the Spitfire, to be replaced by a compound straight tapered wing with a laminar-flow section designated Supermarine 371 I/II. This wing had the maximum thickness at 40 to 42% of the chord and although the wing did not require any special construction methods, the need for a smooth external surface required that production standards and tolerances were of the utmost importance. The new wing also allowed the Spiteful to feature a wide track undercarriage that retracted inwards, thus replacing the outward retracting narrow track units seen on the Spitfire. Even with the later marks of Spitfire it was obvious that the airframe design still had potential, but that of the piston-engine itself had reached its limit. Once jet engine performance surpassed that of the piston-engine and was able to develop and sustain sufficient thrust there was at last the prospect of developing a new breed of fighter aircraft.

Rolls-Royce had commenced jet engine development with the Welland B/23 to be followed by the Derwent B/26. Their third jet engine, designed and built in just five months, would become the Nene RB.41. Thus 'R' denoting Rolls and 'B' Barnoldswick, their factory location; although today the designation RB is accepted as representing 'Rolls By-Pass'. The engine

ment was revised to 4,500lb thrust, the engine was actually able to surpass this requirement quite easily devcloping about 5,000lb of thrust.

Much of the development in jet technology at this time had been achieved by the combination of better design with straight-through combustion chambers aiding efficiency, and the development and production of new high-grade refractory alloys that allowed engines to run at far higher temperatures without any reduction in their life expectancy. Even producing twice the thrust of any engine to date, strangely the Nene saw little use in the nation of its birth, with just Hawker's Sea Hawk and Supermarine's Attacker adopting it. In the USA Pratt & Whitney produced the

The first prototype, TS409, probably photographed at Chilbolton in 1946/47. FAA Museum

had initially been envisaged as producing approximately 4,000lb thrust and was designated RB.40. This new engine was to be used in a 'Spiteful development' fighter design but Joe Smith indicated that he wanted an engine of a smaller diameter producing about 3,000lb thrust, so the engine design was revised to meet these needs and thus became the RB.41, later to be named, as previously stated, the Nene. It was first bench-tested on the 27th October 1944 and although based on the basic layout conceived by Frank Whittle, the engine featured a double-sided centrifugal compressor that gave higher compression and thus greater thrust. It featured nine combustion chambers and a single turbine and although initially it did not reach its estimated output, once intake swirl vanes were added, the engine reached 3,000lb thrust and when the official require-

engine under licence as the J42 powering the Grumman F9F Panther and the Lockheed YP-80A prototype. Later the J42 was provided with reheat becoming the J48 offering 8,000lb thrust. The Nene was also produced for a short time in Australia, where it was fitted to the De Havilland Vampire. Twenty-five Nenes were gifted to Soviet Russia, where they were produced without licence by Klimov as the RD-45 and later in a modified form as the VK-1. This latter version was exported to China by the Russians, where it was again produced without licence as the WP-5!

The Nene was not widely adopted for use in the UK partly because its performance would be surpassed by another Rolls-Royce engine that would later be known as the Avon – a very successful design that would be adopted by many aircraft designers of its era. A further reason for the Nene's lack of application may perhaps be explained by the attitude of the armed forces and Air Ministry. The introduction of the Gloster Meteor into RAF service and the prospect of the De Havilland Vampire to follow shortly afterwards, coupled with the belief in some quarters that the war could end in 1944, perhaps convinced some of the powers that be that there was no immediate need for new designs. This severe lack of foresight prompted a degree of disbelief from the various manufacturers, though fortunately a limited amount of development was eventually sanctioned.

Jet aircraft of the Spiteful type

The birth of the Attacker originated from experimental specification E.1/44, an RAF specification for a jet-powered single-seat fighter. At the time the specification was written the war was still in full fury and orders for both the Spitfire and its replacement, the Spiteful, had not abated. However Joe Smith saw the Nene as an engine with great potential for a jet fighter. Gloster were already working on a completely new design to meet the specification, but Joe Smith knew that the wings and undercarriage of the Spiteful, which were already tooled up for low-volume production, could be switched to the new jet, thus saving a considerable amount of time and money. In fact he estimated it would save six months of development time and as much as £500,000! The new jet thus adopted the Spiteful wings, but as no coolant radiators were needed these were dispensed with and the resulting area was occupied with split flaps and fuel cells. Each Spiteful wing mounted an armament of two 20mm Hispano Mk V cannon, an installation which would be utilised for the Attacker. Retaining the guns in the wings obviously meant that there was no need to incorporate them into the new fuselage thus averting concerns about blast effect on the airflow into the engine intakes. The wing was modified at the root, with a revised stub spar due to the increased diameter of the new fuselage, increasing the overall wingspan to 36ft 11in, and an increase in area from 210 to 226.4sq ft. Whereas certain components of the piston-engined Spiteful/Seafang could and would be used in Joe Smith's fighter, the fuselage of the Spiteful/Seafang could not be. In the mid-1940s little was understood about aerodynamic effects on airframe and engine air flow, or which fuselage shape or configuration would best suit a jet-powered fighter.

The diameter of the Nene engine (49½ inches) determined the minimum diameter of the fuselage. It was mounted in a sealed plenum bay at mid-section, the roof of this bay was detachable to allow access to the engine for servicing and changing. The jet efflux exhausted directly to the rear emerging below the tail. The engine air intakes were situated either side of the fuselage adjacent to the cockpit, a simple 'straight through' arrangement which would later become commonplace but in 1944 was new and many contemporary observers commented on the type's 'ele-

phant ears'! The initial design was submitted to the Directorate of Technical Development (DTD) of the Ministry of Aircraft Production under Supermarine Specification 477 just a few weeks after the first flight of the Spiteful. Allocated Type number 392 it had fuel in five internal cells, equating to a mere 310 Imperial Gallons, there was no fuel in the outer wings at all. Hawker partially solved the lack of internal fuel capacity in the Sea Hawk by using bifurcated exhaust pipes, allowing a fuel tank to be positioned in the rear fuselage, in the 1940s this was a unique design element that Joe Smith obviously did not consider for the Type 392. In the end this lack of fuel would seriously hamper the operational radius of the Attacker, and even though a huge 250 imperial gallon external tank was added, the Attacker never had the range of comparable American fighters of the era. After study of the design, the Ministry of Aircraft Production advised Supermarine that three prototypes of the Type 392 would be required and the order for them was received on the 5th August 1944. The contract for the construction was issued a month later and serial numbers TS409, TS413 and TS416 were allocated to these 'jet aircraft of the Spiteful type' as they were initially called. The first airframe (TS409) was built to a non-naval standard while the second and third were to be navalised but without wing fold mechanism. Once a partial mock-up of the Type 392 was inspected by the Ministry a new specification was written around it: E.10/44.

Development of the new jet aircraft could have been seriously hampered after the loss of the first prototype Spiteful NN660 on the 13th September 1944. However once the second prototype Spiteful NN664 was completed and flown on the 8th January 1945 the jet project was back on track. Initial tests with the Spiteful had shown it to possess poor low-level performance, but this did not deter the Ministry, which with Treasury approval on the 7th July 1945, placed a pre-production order for six E.10/44s (the original RAF requirement) and for a new jet fighter for the Royal Navy, consisting of 18 airframes to Specification E.1/45. The contract issued on the 21st November 1945 allocated serial numbers in the VH980 to VJ118 range (VH980-VH985, VH987-VH990, VH995-VH999 and VJ110-VJ118) to the pre-production batch. Unfortunately the flight of the E.10/44 prototype was delayed while the handling problems experienced with the Spiteful were overcome. One of the major delays was with aileron development, particularly the slotted aileron with which the Spiteful wing was equipped. These latter ailerons were fitted with geared balanced tabs which at speeds over 400 mph became too heavy to operate and as a result the Navy, who were impatient to get a new jet fighter, decided in February 1946 that they no longer wanted the E.1/45 and cancelled their order, buying 18 Sea Vampire F.20s instead. While work on the three prototypes continued the order for the 24 pre-production airframes was subsequently cancelled. Despite this cancellation the second jet prototype, TS413, was allocated for investigation of a handling characteristic known as 'snaking', a directional instability often experienced during the early period of jet development.

First Flight

The first prototype of Supermarine's new jet fighter (TS409) underwent many modifications before being transferred by road to the Aircraft and Armament Experimental Establishment (A&AEE) at Boscombe Down, Wiltshire. It was from there that it made its first flight on the 27th July 1946 in the capable hands of Supermarine's Chief Test Pilot Jeffrey Quill – an exceptional pilot, perhaps best known for his involvement with the testing of the many versions of Spitfire. It performed well, albeit in basic form lacking an ejection seat and other refinements which would be fitted later into production aircraft. It also featured a retractable tail-wheel unit utilising two castoring wheels because of limited space in the aft fuselage. Even before flight testing started it was known that the jet efflux caused damage to both grass and tarmac runways. This problem was never really overcome, but in the end the jet pipe was simply angled up a few degrees to deflect the exhaust gases without resulting in any serious longitudinal trim problems. Initial performance of the prototype was hampered by restrictions in the Nene engine to 12,000rpm, but when the original No.13 engine was replaced with a later one (No.28) the rpm limit was increased to 12,440rpm and as a consequence speed increased from 542mph to 580mph at sea level and 552mph to 568mph at 15,000ft. After initial tests at the A&AEE the type was officially unveiled to the public at the SBAC show at Radlett in September 1946.

Third prototype TS416 fitted with Rocket Assisted Take-Off Gear (RATOG). The photograph is dated 29th September 1950 at which time TS416 was based at RAE Farnborough.
via Tony Buttler

The derelict fuselage of TS416 at Culdrose on 19th July 1958. Accompanying TS416 is Firefly AS.5 VT369 which apparently had been lost over the bows of HMAS Sydney in Lyme Bay on 31st August 1950. The fuselage later being recovered and dumped at this location.
Brian Stainer via Brandon White

Chapter 2: **BIRTH OF THE ATTACKER**

The second prototype, TS413, was modified to meet Specification E.1/45 and designated by Supermarine as the Type 398; it would subsequently be officially named Attacker on the 31st March 1947. Flown for the first time by Mike Lithgow on the 17th June 1947, TS413 differed from the first prototype TS409 in many ways. It had a smaller vertical fin with larger tailplanes, lift spoilers were fitted above the wings and the flaps were also modified. Balanced aileron tabs were added and the air intakes were modified with louvered by-pass bleeds. Also featured was additional fuel in the aft fuselage. The undercarriage was modified to deal with and absorb the stresses created by landing on an aircraft carrier deck. Supermarine avoided the need to revise the undercarriage bays themselves by having the oleo compressed during the retraction cycle to thus shorten it and allow it to fit in the well (something that Republic had done with their P-47). For the first time a Martin-Baker ejection seat was fitted, although Supermarine themselves had tried (unsuccessfully) various ballistic solutions to overcome the problem of ejecting a pilot through the inherent high pressure air of a stricken jet aircraft. Lt P O McDermot RN became the first service pilot to use the ejection seat in an emergency when he ejected from Attacker WA480 on the 20th March 1950.

When TS413 flew for the first time the 'snaking' experienced earlier with TS409 occurred, this time though it was apparent to a much greater degree and over a wider speed range. A length of beading extending from the top of the rudder to the bottom of the trim tab for the second flight largely corrected this problem. During this flight a speed of 375 knots (Mach 0.823) was achieved at 20,000ft. Flight testing was then directed to the lift spoilers and low speed handling, with around 30 airfield dummy deck landings being undertaken. Deck landings on the carrier HMS *Illustrious* commenced in October 1947 with Mike Lithgow and Lt Cdr E Brown of the RAE and Lt S Orr of the A&AEE taking part. The need for thrust spoilers on the exhaust was not considered necessary and even the lift spoilers on the wings were soon abandoned in favour of the adoption of either air brakes made by special extension of the flaps or with specific air brakes built into the fuselage. The tail-wheel undercarriage was in some respects a benefit for deck landing as it induced drag; a tail-down stance was also better for a catapult launch. Sadly TS413 was lost in a crash in June 1948 that claimed the life of its pilot, Lt T J A King-Joyce RN from A&AEE.

The world absolute speed record was by 1947 beyond the Attacker's abilities, as Gloster's Meteor IV had already twice broken and taken it. However a pro-

totype Attacker flown by Mike Lithgow did take the 100km Closed Circuit Record on the 26th February 1948 in TS409 with speeds of 560.634mph and he bettered it on the 27th with a speed of 564.882mph. TS409 was also to take the SBAC Challenge Cup with an average speed of 533mph in July 1950, again with Lithgow at the controls.

Many alternative engine installations for the Attacker were envisaged during these early stages, with Supermarine Specification 510 being written for a projected 'Attacker Mk 2' fitted with a De Havilland Ghost II of 5,000lb thrust and Specification 527, that considered the installation of Rolls-Royce Avon or Tay engines of 6,000lb thrust. Other schemes considered were a two-seat trainer variant and another equipped with floats, but none were developed. Rocket Assisted Take-Off (RATO) was tested on the Attacker, with eight rockets in pairs, two above and two below each wing, but was never adopted operationally. Under Supermarine Specification 512 deflection of the jet exhaust was considered in order to reduce speed on landing, but was never actively pursued.

With the loss of TS413 the Type 392 (TS409) was upgraded to full Type 398 naval specification and flew in this form for the first time on the 5th March 1949. It undertook further development work of the air brake system and additionally to overcome a rudder-locking problem that often occurred when the aircraft was in certain conditions of sideslip. This is believed to have been the cause of the loss of TS413 and may also explain the later loss of production airframe WA477, killing Supermarine test pilot Peter Robarts. Much consideration has been given to the actual cause of the accidents with TS413 and WA477, as at both crash sites Terry-type spanners were found in the wreckage. Although neither crash investigation could pinpoint the spanners as being the cause, it was thought that they could have jammed spoiler operating mechanisms. Whatever the true cause, the rudder-locking was overcome by fitting a dorsal fin to all production airframes, whilst the small number of aircraft already completed would have it retro-fitted.

When prototype number three, TS416, was completed it incorporated all the lessons learnt with the previous two airframes. The wings were moved back by 13 inches and the air intakes were enlarged, both factors that improved the handling markedly.

Into production

The first order for the production Attacker was placed in September 1948 and called for 63 aircraft to Type 398 standard to be designated Attacker F.1. These production aircraft differed from the three prototypes

which had non-folding wings in that they were equipped with the wing fold mechanism as originally required for the Seafang. The first Attacker F.1, WA469, was completed at South Marston and transferred by road to Chilbolton, Hampshire, in March 1950. It was flown for the first time by Mike Lithgow on the 5th April 1950. On the 23rd May 1950 WA469 was to experience a rather unusual incident when conducting a programme of diving and pull outs over South Marston airfield. The pilot, Leslie R Colquhoun, heard a loud bang and looked out to see that the starboard wing tip had folded up to the vertical position and locked there! With full port rudder applied and no ailerons, these being effectively locked by the folded wing tip, Colquhoun made a wide circuit of South Marston and landed at a speed of 210 knots. With the brakes hard on the aircraft suffered a burst tyre and slewed to port, but he was able to bring WA469 to a stop just 30ft short of the runway's end. As a result of his skill and courage, Leslie Colquhoun was awarded the George Medal. WA469 moved to A&AEE Boscombe Down on the 1st August 1950 and was retained there for various trials and elsewhere until eventually arriving at Arbroath, Angus, in September 1955 as Instructional Airframe number A2396. It was struck off charge (SOC) in January 1959.

Attackers WA470-WA484 were accepted into service in a 'sub-standard' condition. Presumably this was to facilitate a rapid introduction of the new jet into naval service, many being used for trials of one form or another. These resulted in the embodiment of various modifications including lighter aileron controls, flat sided elevators and the new dorsal fin previously mentioned. The use of the phrase 'sub-standard' in this context is unusual as opposed to the term 'pre-production' standard. It may be that the term was used to avoid confusion with a batch of 14 E.1/45s (VH995-VH999 and VJ110-VJ118) ordered in 1945 and subsequently cancelled in favour of the Sea Vampire. See page 12.

Right, below, opposite top and centre: *The standard Ministry of Supply set of four views of the first production F.1. WA469. In these photographs the aircraft has the approach angle indicator on the upper nose and radio altimeter 'T' aerial underneath, but lacks the two prominent whip aerials and the second 'T' antenna that would be fitted further aft at a later date.* FAA Museum

The first production F.1, WA469, seen during its time with the Naval Section at A&AEE Boscombe Down. The most obvious feature of this photograph is the unmistakable 250 gallon ventral fuel tank. WA469 at this stage is still fitted with the original small tail. FAA Museum

Opposite top: *A fine study of the first production F.1, WA469, in flight, devoid of any squadron numbers or codes, but with all stencilling in place. No guns are fitted, as confirmed by the 'stubs' in the wing leading edges. The fin extension can clearly be seen and most, if not all, early Attackers built with the original fin had the extension fitted retrospectively.*
FAA Museum

Opposite bottom: *F.1 WA486 with Hispano cannon fitted. Clearly shown in comparison with WA469 (above) is the differing application of the Extra Dark Sea Grey on the top rear fuselage, with the tail fin and rudder finished in Sky.
Whether or not the pilot would have noticed a difference in performance, it seems that the access step has sprung open to an angle of about 45°! via* Tony Buttler

Above: *Built to 'sub-standard' specification, WA472 conducting deck landing trials on HMS* Illustrious *in November 1950. This aircraft was loaned by C Squadron at Boscombe Down to 703 Squadron to undertake these trials and was eventually SOC as an Instructional Airframe in 1953.* FAA Museum

Left: *Prior to squadron allocation F.1 WA497 is seen flying alongside RMS Queen Elizabeth. This aircraft was used by Rolls-Royce for noise suppression trials before serving with Nos.890 and 736 Squadrons. The later style of demarcation can be seen with the Extra Dark Sea Grey extending aft of the vertical tail, which is painted Sky.* FAA Museum

Chapter 3: **INTO SERVICE**

The first jet-powered aircraft to enter naval service were the Vampire and Sea Vampire of various marks. They were acquired from the mid 1940s onwards for experimental and second-line duties. The Vampire's very limited endurance ensuring that it would not have an operational role with the Royal Navy. The Gloster Meteor was also acquired by the Royal Navy for trials and training purposes from the mid 1940s.

The Attacker's place in British aviation history is confirmed by the fact that it was the first *operational* jet to be taken into service by the Royal Navy. The first to receive it was 787 Squadron in January 1951, based at RAF West Raynham, Norfolk; this location was home to the Air Fighting Development Squadron, part of the RAF's Central Fighter Establishment. 787

Above: *F.1 WA484 first issued to 703 Squadron with whom she undertook steam catapult trials aboard HMS* Perseus *in July 1951. In August 1951 WA484 was issued to 800 Squadron becoming 107/J seen here with the early style camouflage finish.* Brian Lowe

Opposite top: *F.1 WA519 111/J of 803 Squadron in 1952.* Brian Lowe

Opposite bottom: *F.1 WA512, 115/J having suffered a late waive off.* via Ian Gazeley

Squadron was termed the Air Support Development Section of the Naval Air Fighting Development Unit. The first Attacker allocated to 787 Squadron was WA478, which arrived on the 6th February 1951 and most of the squadron's initial flying was carried out with this aircraft. It suffered a forced landing at Greenham Common due to engine failure on the 3rd May 1951, but after repair it returned on the 31st August and remained there until being transferred to 890 Squadron at RNAS Ford in April 1952. The squadron also operated WA480, WA486, WA488, WA490 and WA491. WA488 was displayed at Le Bourget on the 26th June 1951 by Lt Cdr Nation (the unit's Commanding Officer) and returned on the 29th June. Sadly WA490 crashed near Cromer on the 6th March 1952 – possibly the pilot suffered oxygen starvation.

The next unit to receive the Attacker was 703 Squadron, which received WA484 in July 1951. The squadron was known as the Service Trials Unit (STU) based at RNAS Ford, Sussex. Amongst many other aircraft types in the STU during the period it operated

the Attacker were: Avro Ansons, Grumman Avenger III and AS.4, Boulton Paul Sea Balliol T.21, Fairey Barracuda TR.III plus various marks of Fairey Firefly and Hawker Sea Fury. This offers some indication of the diversity of work undertaken by this unit.

Although initial trials by 787 and 703 Squadrons had shown the Attacker to be unsatisfactory due to limitations in performance and range, production was too far advanced to be halted. The problem with range was addressed very early on in the Attacker's life with the addition of a 250 gallon ventral fuel tank. Most in-service photographs portray the Attacker with this tank fitted, although its fitment must have been detrimental to the overall performance. Despite known limitations the Attacker was required for service with the fleet, consequently the first operational squadron to receive the type was 800, which formed at RNAS Ford on the 17th August 1951. Their first eight Attacker F.1s were WA473 (102/J), WA484 (107/J), WA487 (108/J), WA492 (104/J), WA493 (106/J), WA494 (105/J), WA496 (101/J) & WA498 (103/J) (squadron codes shown in brackets). Most of the aircraft indicated had arrived on the 22nd August. 800 Squadron was subsequently allocated as a part of the 13th Carrier Air Group that had reformed in 1951 to serve on HMS *Eagle*, which was destined to be the only aircraft carrier to use the Attacker in an operational capacity. HMS *Eagle* had been launched on 19th March 1946 and commissioned for the first time on 1st March 1952. Deck landing practice for 800 Squadron had therefore been conducted on HMS *Illustrious* from late 1951 to February 1952. The squadron embarked on HMS *Eagle* on 4th March 1952 joining 12 Fairey Firebrand TF.5s of 827 Squadron. They remained on board until 24th March 1952, when both squadrons flew back to RNAS Ford. Although 800 and 827 Squadrons were the first operational units to serve aboard HMS *Eagle* it was 703 Squadron that became the first Fleet Air Arm unit to conduct trials aboard the ship, albeit prior to the carrier's first commission.

The next squadron to receive the Attacker was 803, which reformed on the 26th November 1951 and was also allocated to the 13th Carrier Air Group. The squadron was based alongside 800 Squadron at Ford, undertaking a long work-up and conversion to the new type. The squadron complement was eight Attacker F.1s and they were allocated unit codes 111/J to 118/J.

Understandably, training with the Attacker quickly became an important consideration and in March 1952 several aircraft were issued to 702 Squadron at RNAS Culdrose, Cornwall. This was the Naval Jet and Evaluation Training Unit, which operated a mixed

A superb study of F.1 WA486 105/CW from 736 Squadron based at RNAS Culdrose in Cornwall. This photograph, taken off the Cornish coast in 1952 shows the aircraft carrying the early Culdrose station code of 'CW', later changing to 'CU'.
FAA Museum

complement of Sea Vampire F.20s, Meteor T.7s and 10 Attackers (coded 190/CW to 199/CW). The unit was redesignated on the 26th August 1952 following a reorganisation of fighter training units. Number 702 Squadron became 736 Squadron, an advanced jet flying school, and it was given the title Operational Flying School Part II (Jet). The unit operated the Meteor T.7 alongside their Attackers (now coded 100/CW to 119/CW). Due to the work of this unit the Royal Navy had sufficient men to create a third front-line unit.

This final front-line squadron was 890, which commissioned at RNAS Ford on the 21st April 1952 with eight Attacker F.1s (coded 141/J to 148/J). Initially this squadron was intended for embarkation on HMS *Eagle* as part of the 13th Carrier Air Group, but this idea was dropped in favour of the unit becoming a reserve squadron to supply aircraft to the existing 800 and 803 Squadrons.

Number 767 Squadron based at Henstridge in Dorset utilised Attacker F.1 WA470 for a period in January-February 1952 where it undertook Deck Landing Control Officer training. The unit moved to Stretton on the 20th September 1952 and by February 1953 it had three Attacker F.1s and FB.1s as well as a complement of Firefly FR.4s on charge. In Octo-

ber that year it became the Landing Signal Officers Training Squadron as a result of the introduction of the Mirror Landing Aid to Royal Navy aircraft carriers. In July 1953 it received a few FB.2s and continued to operate this mixed bag of Attackers until disbanding on 15th March 1954.

Attacker becomes a Fighter-Bomber

The initial batch of airframes (following WA469), WA470 to WA484, were all completed as 'sub-standard' but it was intended that all Attackers from WA485 would be completed to full F.1 specification. Quite early on in the development of the Attacker however, the Royal Navy had decided that it would be useful to have an offensive bomb-carrying capability. Fortunately the original design had included a bomb-carrying beam within the wing, so all that had to be installed was the associated internal wiring and equipment. Consequently a small number of F.1s were selected for conversion and designated fighter-bomber FB.1. When the first production order for 63 Attackers had originally been placed (WA469 to WA498 and WA505 to WA537) it had been intended, as previously stated, that all would be completed as F.1s, however the last 11, WA525 to WA535, were

Above: *Originally captioned as showing 800 Squadron aircraft ranged on the deck of HMS* Eagle *in Oslo Fjord during 1952. However 111/J in the foreground is WA519 which served in 803 Squadron throughout 1952. This photograph may therefore illustrate one of the occasions when both squadrons were aboard at the same time. Possibly borne out by the fact that at least 12 Attackers are in view, squadron allocation being just eight aircraft until December 1952. Astern is the American heavy cruiser USS* Columbus *CA74.* FAA Museum

Left: *Two F.1s of 800 Squadron at RNAS Ford in early 1952, without station tail code, or carrier code applied.* FAA Museum

converted on the production line to FB.1s. It should be noted here that F.1s WA536 and WA537 were cancelled before production commenced. In addition to the 11 FB.1s mentioned above, five other F.1s were also allocated for conversion on the production line to FB.1s at a later date, (WA481, WA507, WA510, WA511 and WA515). Additionally one other FB.1 was built (WT851) as a replacement for F.1 WA477 that had crashed on 5th February 1951. The first FB.1 to fly was WA526 on 25th August 1951, subsequently being allocated to 703 Squadron at RNAS Ford on 25th January 1952, the first squadron to receive an FB.1 into service.

Of the three operational Attacker squadrons, 800 Squadron was the first to receive the FB.1 in February 1952 and although 890 Squadron received a few FB.1s, none reached 803 Squadron. Both 800 and 803 Squadrons embarked HMS *Eagle* from 4th June to the 10th July 1952. Number 803 Squadron again embarked HMS *Eagle* for Exercise *Mainbrace* on 3rd September 1952 and was joined by 800 Squadron a day later. These two squadrons remained on board until the 9th and 11th October respectively, when they returned to RNAS Ford.

The FB.2

Eighty-four new build FB.2s were ordered, differing from the FB.1 by virtue of the availability of the Nene 7 engine (later redesignated Nene 102) producing 5,100lbs of thrust. Also provided were: electric starter, high energy igniter system, acceleration control linked to the throttle, metal-framed canopy and provision for 12 rockets to be carried, (six per wing mounted in two tiers) plus a small increase in fuel capacity. Additionally FB.1 WA507 was also modified to FB.2 standard. The first FB.2 was WK319 flying for the first time on the 25th April 1952.

The first unit to receive the Attacker FB.2 was 890 Squadron, eight arriving at Ford in July 1952. The squadron reached its full strength of eight operational aircraft plus reserves by October with all earlier marks being passed to other units. 890 Squadron then deployed to Milltown, Morayshire, on the 27th October prior to embarking HMS *Eagle* on the 29th. Number 800 Squadron was the next to receive the FB.2, with WK331 arriving on the 10th September 1952 to begin the process of replacing its FB.1s, although as stated earlier, 800 Squadron was on board HMS *Eagle* at this time thus it begs the question, did re-equipment to the new mark occur whilst embarked?

Opposite top: *An undated photograph aboard HMS Eagle with a mixed complement of F.1 Attackers, Avengers and a single Firefly. The Avengers display an earlier style of roundel.* via Ian Gazeley

Opposite bottom: *Attackers of 800 and 803 Squadron ranged on the deck of HMS Eagle with AEW Skyraiders and Fairey Fireflies in the background.* FAA Museum

Below: *With the Forth Bridge and an escorting destroyer in the background, HMS Eagle leaves Rosyth on an unknown date. On deck are Attacker F.1s with Blackburn Firebrands on the ships stern.* via Ian Gazeley

Above: *FB.1 WA533
served with 800 and 890
Squadrons before joining
736. Coded 115/CU in
September 1953 then
115/LM in the following
November. Possibly taken
on 17th May 1954
following an accident after
landing aboard HMS
Illustrious when both port
flaps were damaged.*
Brian Lowe

Right: *Attackers of 800
Squadron being prepared
on the deck of HMS Eagle
during exercise Mariner.*
FAA Museum

Opposite top: *Superb
atmospheric image of 803
Squadron FB.2s, with
possibly WP277 in the
foreground, being refuelled
and rearmed. The paper
discs seen fitted to the
muzzles of the inboard
guns indicate that these
weapons are being
rearmed.* FAA Museum

Opposite bottom: *FB.2,
WP300, 151 of 803
Squadron on HMS Eagle is
seen being brought up to
the port catapult. The
bridle can be seen hung
over the 'T' antenna! This
aircraft ultimately crashed
short of the runway at Hal
Far, Malta, in April 1954
and was struck off charge.*
FAA Museum

By 9th October and the 11th respectively 803 and 800 Squadrons were back at Ford following their deployment aboard HMS Eagle. However both squadrons rejoined their ship on the 7th November 1952 (803 Squadron still equipped with the F.1 variant at this time) where they joined 890 Squadron. While aboard Eagle, the Attackers of 800 Squadron took part in a fly-past for Her Majesty The Queen at Lee-on-Solent on the 21st November, all the squadrons returning to Ford on the 3rd December, where 803 Squadron re-equipped with FB.2s.

With just eight aircraft (plus reserves) allocated to each operational squadron, the Royal Navy considered this to be too small a force to be effective. As a result 890 Squadron was disbanded on its return to Ford, thus allowing its aircraft and aircrew to be absorbed by the remaining two squadrons. 800 and 803 Squadrons consequently expanded to an operational strength of twelve aircraft each, 803 Squadron for instance having 15 aircraft on strength (presumably including reserves) by early 1953.

Throughout 1953 and the first half of 1954 the two remaining operational squadrons embarked HMS Eagle on several occasions, for voyages of varying duration in home waters and the Mediterranean. Both squadrons in concert with 736 Squadron took their place in the Coronation Review at Spithead on 15th June 1953.

By the beginning of 1954 the Attacker had been deemed obsolete and its replacement, the Hawker Sea Hawk, was already entering service. As a consequence in June 1954, 800 Squadron disbanded, its aircraft being ferried for storage at Aircraft Holdings Unit, RNAS Abbotsinch, Renfrewshire, or to Airwork Limited at Gatwick. There they would be overhauled, some later to be issued to the Royal Navy Volunteer Reserve (RNVR) squadrons from May 1955. Number 803 Squadron left Malta to return to the UK aboard HMS Eagle in June 1954, and on arrival in the UK the squadron was progressively re-equipped with the Sea Hawk FB.3 and the displaced Attackers were sent for storage.

Right: *Most probably FB.2 WZ282 105/J of 800 Squadron and believed to be at RNAS Ford, if the 'FD' stencil on the rear of the trolley accumulator is to be believed. The date would be during the second half of 1953.*

Below: *FB.2 WK322 102/J of 800 Squadron possibly on 4th March 1953, being disentangled from the nylon crash barrier, looking almost undamaged. See production history.*

Opposite top and bottom: *Two views of FB.2 WK331 dated August 1952. As yet devoid of unit markings prior to delivery to the Royal Navy on 20th August 1952. All Brian Lowe*

VICKERS ATTACKER F.B. MK
(Production)
NENE.
AUG 1952.

This page top, below and opposite page top: *Attacker F.1 WA498 103/J having had its arrester hook pulled out whilst landing, continued at high speed into the jet-barrier with predictable results. Published records indicate the date to have been 21st September 1952. However these photographs taken from a contemporary album, are dated October 1952 and state that WA498 was the 'CO's bus'. Whatever the date, unsurprisingly, repairs were deemed unviable and the aircraft was scrapped.* via Ian Gazeley

Opposite bottom: *On 1st April 1954, FB.2 WK332 106/J of 800 Squadron, was only able to lower its starboard undercarriage whilst landing aboard Eagle, WK332 bounced over the arrester wires and into the jet-barrier causing the port wing to be torn off. The aircraft was not repaired.* FAA Museum

HMS EAGLE – The only carrier to operate the Attacker

Eagle was launched in 1946, completed in October 1951, commenced builder's trials during the same month and was commissioned in early 1952. As completed, she was in some respects similar to the six armoured fleet carriers of the *Illustrious/Implacable* class on strength with the Royal Navy at the termination of the Second World War. *Eagle*'s displacement of 36,800 tons (standard displacement) was over 10,000 tons heavier than *Illustrious* (standard displacement) and was approximately 60 feet longer in overall length than *Illustrious*; *Eagle* measuring 803 feet 9 inches overall in 1952. *Eagle* was completed with an axial flight deck, a feature that would be altered with the advent of later naval jet aircraft to incorporate an angled flight deck, although not during the lifetime of the Attacker's association with *Eagle*. In fact *Eagle* would be taken out of service in mid 1954 for a refit which, when completed, would incorporate an interim 5 degree angled flight deck (some sources state 5½ degrees), that required the decom-

missioning of her forward port side turrets, the crowns of which formed, in practice, part of the flight deck. As originally built *Eagle*'s defensive armament consisted of 16 x 4.5 inch guns mounted in eight turrets, sited in pairs on each quarter. Additionally, reflecting perhaps the allied naval experiences with Kamikazes in the Pacific in 1944/45, *Eagle* was equipped with 61 x 40mm Bofors guns, comprising eight six barrelled mounts, two twin and nine single.

Some sources indicate that in service HMS *Eagle* could achieve a maximum speed of 31½ knots when producing 152,000 shaft horse power, other sources state a more credible 30½ knots. At deep load displacement – a little over 50,000 tons, she could achieve 29½ knots. As originally designed she would be able to accommodate 80 aircraft, but this quantity reflects a Second World War requirement and by the 1950s with aircraft growing in size, her complement of aircraft would be significantly less, certainly in peacetime. HMS *Eagle* remained in service, undergoing several refits, finally to be retired in early 1972 and retained as a source of spare parts for HMS *Ark Royal*. *Eagle* was finally broken up in late 1978.

This photograph illustrates the pleasing lines of HMS Eagle in her original form. Arrayed on deck are Attackers, De Havilland Sea Hornets, Fireflies and Skyraiders. Trailing in Eagle's wake is a fast minelayer; Manxman, Apollo or Ariadne.
FAA Museum

An excellent study of HMS Eagle, showing her original deck form and gun turrets to great advantage, being nudged into her berth at Toulon on 4th April 1954 for a short visit before sailing again on April 6th, bound for Naples, conducting exercises en-route. During one such exercise an 803 Squadron Attacker, flown by a Lt Ward, managed to stop his aircraft using brakes alone, having pulled his hook out whilst landing aboard. As a result, Ward's wingman was forced to divert temporarily to a French airfield until the ship's arrester system could be repaired.

On Eagle's port beam is a French aircraft carrier, either La Fayette (R96) or Bois Belleau (R97); possibly the former as the latter was deployed to Indochina in 1954. Both were ex-US light aircraft carriers, having previously been named USS Langley (CVL 27) and USS Belleau Wood (CVL 24) respectively, and were veterans of the war against Japan having been involved in many of the Pacific campaigns. In fact, whilst in French naval service, both ships carried air groups very similar to those carried by these and their seven sister ships in US service during the previous decade in the Pacific, comprising (usually) 24 Grumman Hellcats and eight Curtiss Helldivers, although by 1954 these were being supplemented or replaced by the Vought F4U Corsair and Grumman Avenger in the French Navy. Very evidently, whichever of the two ships this is, the complement of lighter coloured aircraft seen on deck are of a later vintage than the Hellcats and Avengers previously described and are probably Dassault Ouragans and presumably the ship is being used as an aircraft ferry. Other identifiable aircraft types on board include: F4U Corsairs, Helldivers, Avengers and possibly Hellcats.

On Eagles starboard beam are two French light cruisers of the 'La Galissonniere' class, possibly Gloire and Montcalm, two of three ships from a class originally consisting of six, to have survived the Second World War, (the third survivor being George Leygues). All three were completed in 1937, mounting nine six inch guns in triple turrets. All three saw post war service and were condemned in 1958 and 1959 and subsequently scrapped, although Montcalm was renamed as hulk Q457 and released for scrapping as late as the very last day of 1969.

The warship moored to the extreme left of the photograph is a Daring class destroyer, of which eight were built for the Royal Navy. They were the largest conventional destroyers constructed for the Royal Navy and the last of the classic destroyer designs to enter service. The identity of this particular vessel is not known.
FAA Museum

Chapter 4: RNVR, SECOND-LINE AND FOREIGN SERVICE

FB.2, WK320, 833 was operated by 1833 RNVR Squadron and is seen here at RAF Honiley on 26th August 1956. The squadron formed part of the Midlands Air Division. Newark Air Museum

Reserve Squadron Service

By August 1954 the two operational RN squadrons had relinquished the Attacker and most second-line units followed suit during the course of the year. Once considered obsolete it was decided that the Attacker should be issued to the RN Volunteer Reserve (RNVR) squadrons. Consequently on 25th April 1955 718 Squadron was formed at Stretton, Cheshire. This unit undertook pilot training of those RNVR squadrons chosen to operate the Attacker. These were 1831, 1832 and 1833 Squadrons, all of which were operating the piston-engined Hawker Sea Fury FB.11 at that time. Although the Attacker was, by now, itself regarded as obsolete it still offered a significant increase in performance over the Sea Fury. Initial jet training was undertaken in Sea Vampire T.22s, and type conversion completed on one of their fourteen Attacker FB.2s. 1831 RNVR Squadron from the Northern Air Division was the first to convert to the Attacker, in early May 1955. By June they were fully converted having received a total of seven FB.2s, obtained partly by transfer from 718 Squadron, plus others from storage (for example WZ289) to reach their full complement. Next to convert was 1833 Squadron at Bramcote, Warwickshire, within the Midland Air Division. Bramcote proved unsuitable for jet operations, so 718 Squadron having moved south from Stretton relocated to nearby Honiley in July 1955. By October they had completed the conversion of 1833 Squadron and 718 disbanded in December. The final unit to convert to the Attacker was 1832 Squadron at Benson, Oxfordshire, as part of the Southern Air Division. No evidence exists to show that 718 Squadron assisted this unit in converting to the type, although this may well have been the case. During September-October 1956, 1832 Squadron relinquished their Attackers in favour of the Sea Hawk F.1.

The Attacker remained with 1831 and 1833 until the 10th March 1957 when all RAFVR and RNVR squadrons were disbanded as a result of cuts in defence expenditure.

Most ex-RNVR Attackers were placed in storage at the Aircraft Holding Unit (AHU) at Lossiemouth, but as there was no intended future use for them they were sold for scrap during April-May 1958.

Second-line Service

Several Fleet Air Arm squadrons continued to operate the Attacker in a second-line supporting roll.

Number 703 Squadron had been involved in trials work since 1945 and in 1948 moved to Lee-on-Solent becoming the Service Trials Unit, acquiring that role from 778 Squadron. It moved to Ford in April 1950 and received its first Attacker F.1s in July 1951, although these only remained there for a month before departing in August. The Attacker was reintroduced to 703 Squadron in January 1952 in the form of the FB.1. These departed in April that year, the Attacker not returning to the squadron until April 1955, when the FB.2 variant was received. Here various handling trials were undertaken on the carriers HMS *Eagle* in April 1955 and HMS *Ark Royal* in June-July 1955. The unit ceased to exist on the 17th August 1955 when it was amalgamated with 771 Squadron to form 700 Squadron on the following day. Although 700 Squadron is listed as having operated the Attacker, only one example, FB.2 WZ277 is known to have been allocated, 700 Squadron may well have used others but their identities remain unknown. Certainly WZ277 was on 703 Squadron's strength on disbandment, before appearing on 700 Squadron's books the following day and remaining with the latter unit until 15th February 1956, when it went to the AHU at Abbotsinch. Its role within the squadron is unknown.

736 Squadron moved from Culdrose in Cornwall (see chapter 3) to Lossiemouth, Morayshire, on the 9th November 1953, as a part of the Naval Air Fighter School. This relocation was considered necessary because the weather conditions in the north of Scotland proved much more suitable for the flying training needs of this unit. In May 1954 it expanded its Attacker complement to 28 aircraft comprising F.1s FB.1s and FB.2s, but by August these were all withdrawn when the squadron re-equipped with the Sea Hawk.

As outlined at the beginning of chapter 3, 787 Squadron was located at RAF West Raynham, where it had been based since moving there in November 1945. As the Naval Air Fighting Development Unit it undertook trials of various items of equipment and

This page and opposite: A series of photographs covering an exercise involving 1833 Squadron from August 3rd to August 17th 1956 at Ford. The exercise required that 11 Attackers and a single Vampire be flown from Honiley to Ford for the duration. The Attackers were divided into two flights one for air-to-air firing and the other for air-to-ground attacks using rockets (RPs). The squadron experienced problems with the RPs as several pilots suffered 'hang-ups' caused by electrical leads known as 'pig tails' coming adrift in flight. This problem was overcome and results improved with each pilot flying approximately 10 RP sorties. (cont. page 37)

weaponry for Royal Navy operations, using the Attacker F.1 from February 1951 to April 1952, the FB.1 from September 1952 to September 1954 and the FB.2 from April 1954 to September 1954. In conjunction with the Attacker FB.1 and FB.2 the unit also operated Sea Hawks and Westland Wyverns to conduct these trials, the squadron carrying on with this work until 16th January 1956 when it disbanded, by which time the unit was utilising the DH Sea Venom FAW.21.

Civilian Service

Attackers also saw use with the Fleet Requirements Units (FRU), run by Airwork Limited for the Royal Navy at Hurn, Hampshire – later Dorset, and Brawdy, Pembrokeshire, and its satellite airfield, St. Davids, Pembrokeshire. Initially it had been thought that the Fairey Firefly would replace the DH Sea Hornet in FRU service, but this was discarded in favour of using the Attacker. The idea was to use the Attacker on a trials basis with the intention of assuming the gunnery and tracking role undertaken at this time by 700 Squadron. The Hawker Sea Fury, of which eight were allocated covered all the other general FRU duties.

However, as the Sea Hornets left the FRU before the first Attackers arrived, the Sea Fury contingent for a time, undertook all the roles within the unit. Regarding operations from Hurn, there were initial concerns not only with a civilian company operating a jet-powered military aircraft, but also because they were operating from a civil airport. Six Attackers were delivered to Hurn, the first WZ286, arrived on the 1st November 1955, and the last, WP285, on the 18th January 1956. At the same time a mixed bag of eight F.1s, FB.1s and FB.2s were supplied to Brawdy/St. Davids. The Attacker was always regarded by the FRU as an interim type, and as such was soon replaced in the FRU at Hurn by the Sea Hawk F.1 from November 1956 and the Sea Venom at Brawdy/St. Davids from February 1957. The last airworthy Attacker FB.2 WZ286 departed Hurn on the 26th February 1957, their very last Attacker FB.2 WP297 left by road transport to RNAY Fleetlands on the 15th March 1957. The last Attacker to leave Brawdy/St. Davids was F.1 WA513, in July 1957. Most of the ex-FRU aircraft were subsequently placed in storage at AHU Abbotsinch, and sold for scrap during 1958.

The air-to-air firing programme was less successful. Each aircraft was armed with 60 rounds per gun, but problems were experienced with the target banners towed by Sea Hawks of 700 Squadron, only one banner being retrieved intact out of 16 flown. FAA Museum

The following statistics were listed at the conclusion of the exercise:

Sorties flown	233
Average serviceability	85%
Hours flown	not recorded
Rounds fired	5,789
Rockets fired	269
Banners taken off	16
Banners returned	1 (hits nil)

Right: *FB.2 WZ273 823 the first of her production batch seen with 1832 Squadron and bearing evidence of earlier times with 736 Squadron judging by the remnants of red paint at the tip of the nose. The choc marked 'SAD' in the foreground refers to Southern Air Division.* Brian Lowe

Centre right: *Taken from the 1833 Squadron archive showing damage caused to the fuselage of one of their Attackers. Sadly the date, and aircraft identification remain unknown but the following quote was found attached –*

'Having congratulated his Pilot's Mate on the smooth running and general excellence of his engine, Nigel was dismayed to find this hole in the fuselage as he climbed out of the cockpit. The flames from the exploded flame tube touched the fin and rudder, but fortunately no further damage had been done'. FAA Museum

Below: *An 1833 Squadron Attacker lies wrecked, at Honiley. This photograph from the squadron record book was accompanied with the caption 'all done by mirrors'. It would seem that in following his reflection in the deck landing mirror the pilot flew a little too low and collected some shrubbery before hitting the runway rather hard! The gap in the boundary hedge can be seen to the right of the RAF ambulance in the background.* FAA Museum

Opposite top and bottom: *Busy scenes at Honiley. The removal of the nose cone (bottom right) reveals amongst other things the large and heavy lump of lead required to maintain the Attacker's centre of gravity.* FAA Museum

Above: *Attackers of 736 Squadron undertaking formation flights on Thursday 4th March 1954. 736 served as an advanced jet flying school operating the Attacker F.1 and FB.2 until August 1954, by this time based at Lossiemouth.*
FAA Museum

Right: *'Sub-standard' specification early production F.1 WA482, which became Instructional Airframe A2407 in 1956 and seen here with the Air Engineering School at Arbroath. It was SOC there in January 1959.*
FAA Museum

Above: *FB.2 WZ300 836 following a landing accident after overshooting the runway at Ford on 10th August 1956. A clue as to why this happened may be gleaned from the following 'ditty' that accompanied the photograph in the 1833 Squadron archive. WZ300 was subsequently broken up for spares as a result of this accident.*

Doc Greenlaw went rocketing at Bracklesham Bay,
But alas a great thunderstorm got in the way,
He came down still loaded and too bloody fast,
And Commander Air's fence was a thing of the past,
Cracking show, he's alive,
But he still has to render his A.25.

FB.2 WZ292 838 of 1833 Squadron at Honiley following an accident in late 1956. Having taken off the aircraft failed to climb as the lift spoilers were jammed, the pilot retracted the undercarriage and the aircraft continued through the airfield boundary, across a road and into a field on its belly. Close examination of the photograph reveals that the underside is supported by a wheeled trolley. WZ292 was not repaired.
Both FAA Museum

Attackers with the Royal Pakistan Air Force

Pakistan was the only export customer for the Attacker. In the late 1940s and early 1950s there were numerous international embargoes that prevented the sale of weapons and associated material to many developing nations and as a result there was little opportunity to find extensive sales worldwide. Just 36 airframes were ordered by Pakistan and they were basically de-navalised FB.2s that retained the capacity to carry bombs and rockets but had no wing-fold. Each was equipped to carry the 250 gallon ventral fuel tank and at least one photograph exists showing a Pakistani Air Force Attacker equipped with a 93 gallon drop tank beneath each wing, in addition to the ventral tank. Whether or not the drop tanks were ever used is not known, but certainly the RPAF used the Attacker operationally with the ventral tank. It is perhaps possible that the additional drop tanks were just used during the ferry flights. The first 33 aircraft were ferried to Pakistan by Supermarine pilots. The last three were ferried by pilots of the Royal Pakistan Air Force using the event as an exercise logging an average flying time of 11hr 40min.

Prior to delivery each aircraft had been test-flown in the UK bearing their RPAF serial numbers (R4000-

being allocated on a consecutive basis – the last, R4035, being identified as G-15-211. Oddly, R4000 was identified out of sequence as G-15-110. The letter 'G' of this registration represents the nationality, in this case Great Britain, the number '15' identifies the company, Supermarine; the final group of numbers identify an individual aeroplane from that company.

The first and only RPAF unit to operate the type was 11 'Arrows' Squadron, which had been formed as a light bomber unit on 1st January 1949 at RPAF Station Mauripur. The intention was to have equipped this unit with the Bristol Brigand, however the first Brigand crashed on its way to Pakistan and procurement of the rest was subsequently cancelled and the squadron de-activated. In June 1951 the unit was reactivated as a fighter/interception squadron under its first squadron commander, Squadron Leader A Rahim Khan. Its first Attackers arrived at about this time and it remained the only jet squadron in the Royal Pakistan Air Force. On the 18th January 1956, 'Arrows' Squadron began to re-equip with the North American F-86F Sabre, supplied under an American military aid programme (MAP) and became a fighter-bomber squadron. The Attackers are then believed to have been placed in store, though no details are to hand concerning their fate. The RPAF experienced

Royal Pakistan Air Force Attacker R4033 is seen here at Chilbolton airfield in the company of Royal Navy Attackers prior to delivery. The British identification number G-15-209 was applied for test flights in the UK. Authors Collection

R4035) as well as a temporary identification often referred to as 'B conditions', 'Class B (Provisional) Registration' or sometimes 'G class registrations'. Certainly official references in the Air Navigation Order and the British Civil Airworthiness Requirements of the day clearly states that aircraft flown under 'B conditions' are deemed as having been identified as opposed to registered. As a consequence R4001 was identified as G-15-177 with subsequent numbers

frequent attrition of the Attacker caused mainly by maintenance and technical problems. The undercarriage was particularly prone to damage or failure, which in itself caused the loss of a number of aircraft.

The Attacker had left RPAF service prior to Pakistan becoming an Islamic Republic within the British Commonwealth on 23rd March 1956. At this point the 'Royal' prefix was dropped by the RPAF.

Above: *An unobstructed view of R4033 with Meteor F.8 in the background.* Andy Thomas

R4002 at RAF Shaibah near Basrah, Iraq, en-route to Pakistan. Andy Thomas

R4002 and R4003 at RAF Habbaniya, Iraq, en-route to Pakistan. Andy Thomas

R4017 and R4016 prior to or during delivery at an unknown location. Tony Buttler

Chapter 5: **F.1 TECHNICAL DESCRIPTION AND ARMAMENT**

The Attacker was a single-engined, jet-propelled, single-seat monoplane of stressed skin construction with retractable tail and main wheels and folding mainplanes. The complete airframe description is as follows.

Fuselage
The fuselage was of circular section, except for the nose portion, and was built up of bulkheads, frames and intercostal members which supported a light alloy skin. The nose portion resembled a flattened 'O' when viewed from the front; where the section abruptly changed to circular, the chords thus formed were utilised for the intake of air into the engine. The nose portion accommodated the pilot in a cockpit which could be pressurised and housed the radio and the electrical accumulators.

Immediately aft of the cockpit was the fuel tank bay, where the centre and the side tanks occupied the spaces on either side of, and between, the converging air intakes. The bulkhead that terminated the rear of the tank bay provided attachment for the mainplane front spars and formed the division between the tank bay and the engine bay.

The bulkhead at the rear of the engine bay coincided, in position, with the trailing edge of the wing root fillet and from this point the fuselage tapered to the tail end where the jet pipe emerged.

Wings
The laminar-flow type mainplanes were of stressed skin construction with main and auxiliary spars and ribs that supported a heavy gauge skin. Detachable wing tips were fitted, and the mainplanes could be folded under hydraulic power. Slotted ailerons were employed and each had a spring tab; the starboard aileron had, in addition, an electrically-operated trim tab. The mainplanes were attached directly to the fuselage, no centre-section being incorporated in the fuselage. A pressure-head was fitted at the leading edge of the port wing tip.

Tail
The tailplanes, elevators, fin and rudder were of all-metal construction. The tailplanes were conventional in design and location and each had a 10° dihedral. The elevators were horn-balanced and each had a trim tab; in addition the port elevator was fitted with a spring tab. The fin and rudder was mounted centrally on the fuselage, the trailing edge of the rudder coincided with the leading edge of the elevators. The rudder had an electrically-operated trim tab.

Undercarriage
The undercarriage consisted of two separate main wheel units and a tailwheel. The main wheels retracted inwards and upwards in the mainplanes where they are completely enclosed by doors. The tailwheel unit, which had twin wheels, retracted aft into the fuselage and was enclosed by doors. The main wheel units had long-stroke oleo-pneumatic shock absorber struts fitted with pre-retraction gear, and pneumatic brakes were employed. The tailwheel unit was fully castoring, had a self-centring device and could be steered and locked.

Engine
The engine was attached rigidly to the fuselage at three points, the two side attachments being directly to one of the fuselage frames and the forward attachment being made by a strut. The engine was started electrically.

AUXILIARY SYSTEMS

Flying Controls
The flying controls had a conventional rudder bar, stick-type control column and the rudder pedals could be adjusted for leg reach. Lateral movement of the control handle at the top of the control column operated the ailerons through cables and rods, and cables were used to control the rudder and elevators. Mechanically linked inner and outer trailing edge flaps were fitted and were operated hydraulically, as also were the wing-lift spoilers which were controlled by a pre-selector valve and used as air brakes in conjunction with the flaps.

Electrical System
The electrical system derived its power from a 30 volt, 100 amp, engine-driven generator charging two series-connected 12 volt, 40 amp batteries. The electrical services included the fuel pumps, fire extinguisher system, gun firing, ciné camera operation, instruments, indicators and warning lamps, ASI, pressure-head, ailerons and rudder trim-tab actuators and engine starting. Provision was made for wireless and radar installations as follows:

Wireless	Radar
ARI.5272	ARI.5679 IFF
TR.1520 radio	ARI.5284 radio
TR.1936 VHF radio	altimeter
ARI.5307 ZB/X beacon	
homing	
A.1271 beam approach	

Fuel

The fuel supply was carried in seven tanks, five of these were in the fuselage and two in the mainplanes. Provision was also made for carrying a drop tank beneath the fuselage. The tanks were of the crash-proof flexible type with the exception of the fuselage centre tank, which was of rigid construction. Fuel was transferred by air pressure, the rate of transfer being controlled automatically.

The fuel tank capacity was as follows:

Fuselage centre tank	082 gallons
Fuselage port side tank	073 gallons
Fuselage starboard side tank	073 gallons
Fuselage port rear tank	012 gallons
Fuselage starboard rear tank	012 gallons
Inter-spar port wing tank	20½ gallons
Inter-spar starboard wing tank	20½ gallons
Maximum internal fuel capacity	293 gallons
External ventral tank	250 gallons

Total 543 gallons

All the following drawings are taken from the official maintenance manual and are Crown Copyright.

Attacker F.1 cutaway.

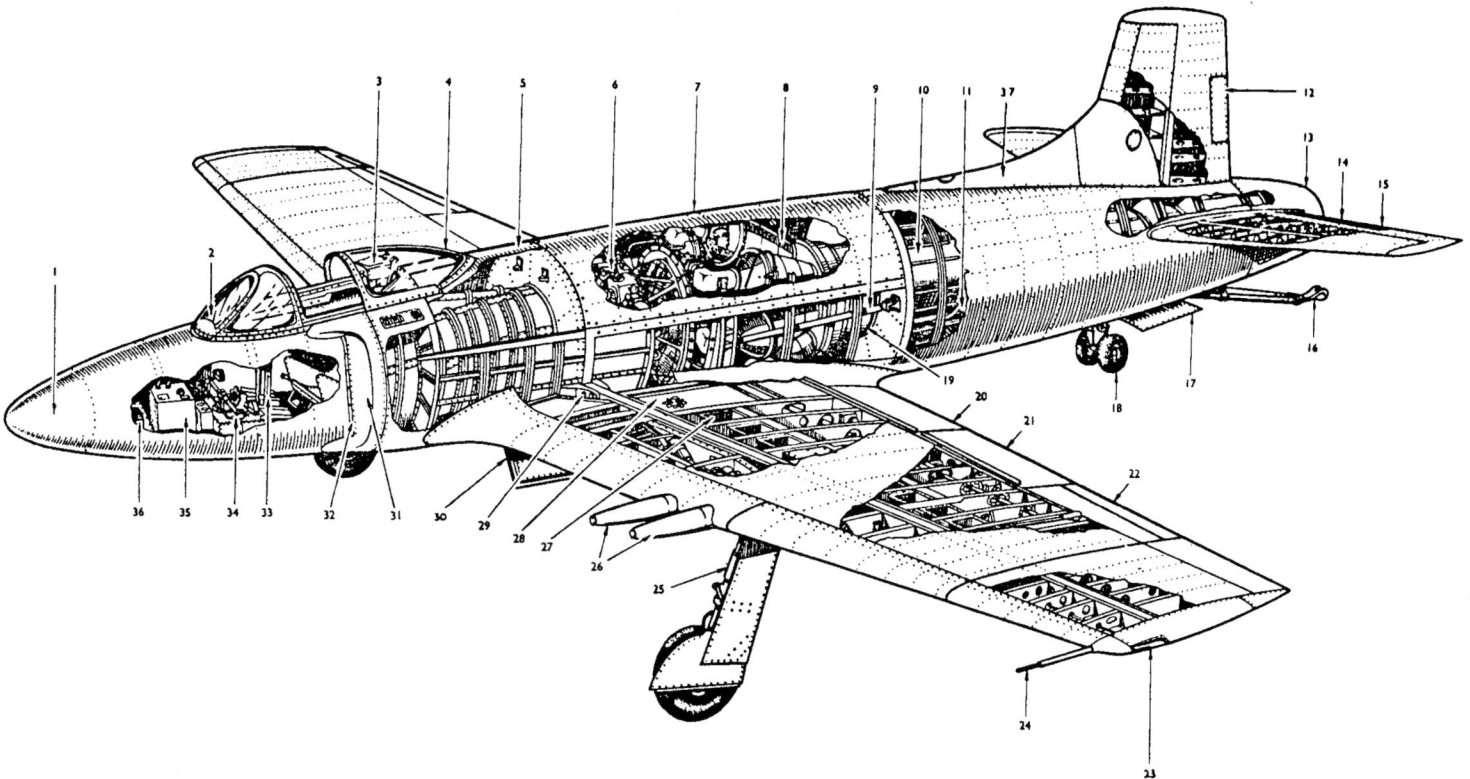

1 Nose cap	15 Elevator trim tab	29 Wing main attachment to fuselage
2 Glass fairing and B.P. windscreen	16 Arrester hook	30 Main wheel door
3 Ejector seat	17 Tail wheel doors	31 Air intake
4 Sliding hood	18 Steerable twin tail wheels	32 Boundary lever bleed
5 Tank bay cover	19 Engine bay air bleed	33 Control column
6 Accessory drive gearbox	20 Inner trailing edge flap	34 Rudder bar
7 Engine bay cover	21 Outer trailing edge flap	35 Radio
8 Nene 3 engine	22 Aileron spring tab	36 Electrical accumulators
9 Engine 'crash' strut	23 Navigation lights	37 Dorsal fin
10 Jet tube	24 Pressure head	
11 Exhaust valute	25 Shock absorber strut with fairings	
12 Rudder trim tab	26 Hispano 20mm guns	
13 Rear fairing cone	27 Ammunition bays	
14 Elevator spring tab	28 Wing fuel tank	

Oil

Oil was carried in the engine sump only.

Pneumatic System

The pneumatic system operated the wheel brakes with compressed air from a storage cylinder, which was maintained by an engine-driven compressor.

Hydraulic System

The hydraulic system operated the alighting gear and tailwheel steering, the fairing doors on the wheel units, the arresting hook, the wing-lift spoilers and the flaps. The system was of the live-line high pressure type and derived its power from an engine-driven pump unit that embodied a two-stage pump and a cut-out; a hand pump was also fitted. An emergency system was provided to lower the alighting gear in the event of failure of the hydraulic system. The total capacity of the hydraulic system was 3 gallons 1⅔ pints

General Equipment

General equipment includes an oxygen system, cockpit pressurisation, windscreen de-icing and de-misting, provision for an anti-G suit, deck arresting gear, deck assisted take-off gear and RATO. The aircraft was fitted with a Martin-Baker Mk 2A ejection seat.

Armament

Armament consisted of four 20mm Hispano No.3 Mk 5* cannon, two in each mainplane, provided with 624 rounds of ammunition (167 rounds each for the two inboard and 145 rounds each for the outboard weapons). Aircraft that embodied Mod.124, 125, 126 and 127, i.e., the FB version, were equipped to carry, in addition to the fixed armament, 3in unguided rocket projectiles and bombs.

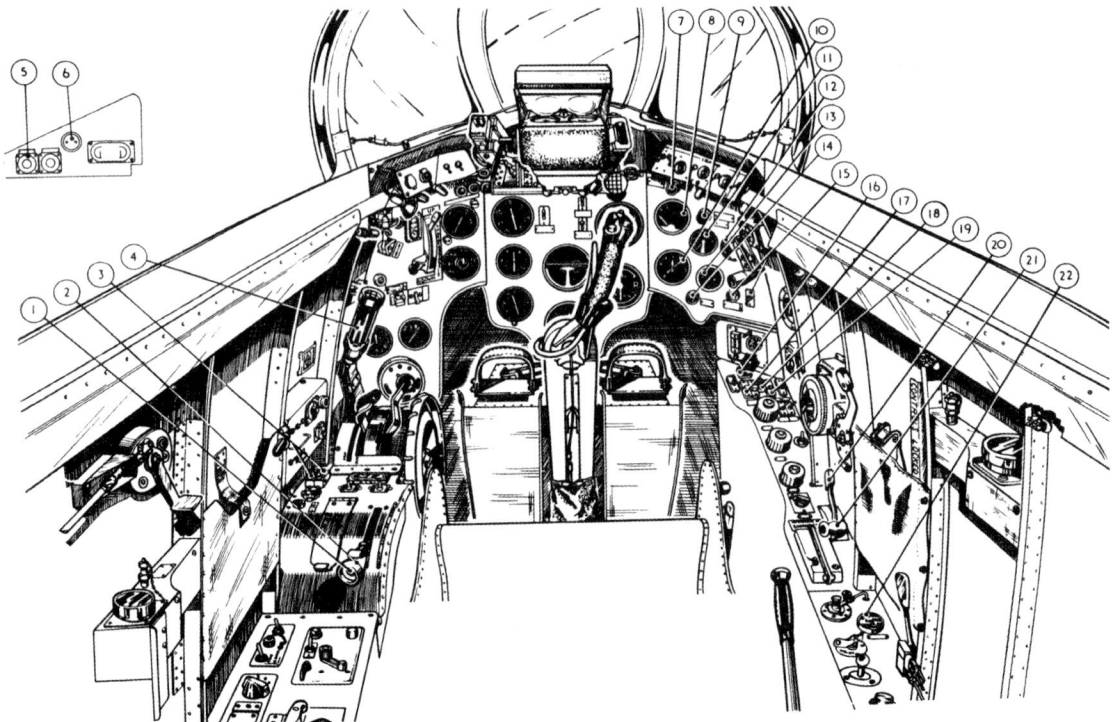

F.1 engine controls and instruments.

1 Engine relight switch	10 Tachometer	20 Fuel transfer cock lever
2 High pressure fuel cock lever	11 Main tank contents gauge	Drop tank - Forward Internal tank - Aft
Open - Forward Closed - Aft	12 Oil pressure gauge	21 Low pressure fuel cock lever
3 Fuel Pump Circuit Breakers	13 Auxiliary tanks contents gauge	On - Forward Off - Aft
4 Throttle control lever	14 Fuel pressure warning lamp	Lever must not be off when engine is
Open - Forward Shut - Aft	Red when pressure drops below 3psi	turning
5 Fuel pump test button	15 Fuel pump isolating switch and warning	22 Fuel transfer pressure gauge
6 Fuel pump test socket	lamp. Red when fuel pumps are isolated	
7 Generator failure warning lamp	16 Engine starting switch	
8 Jet pipe temperature gauge	17 Engine starting safety switch	
9 Fuel transfer warning lamp	18 Generator field circuit breaker	
red when there is no fuel transfer	19 Ignition Isolating switch. For use when	
	turning the engine	

F.1 access panels.

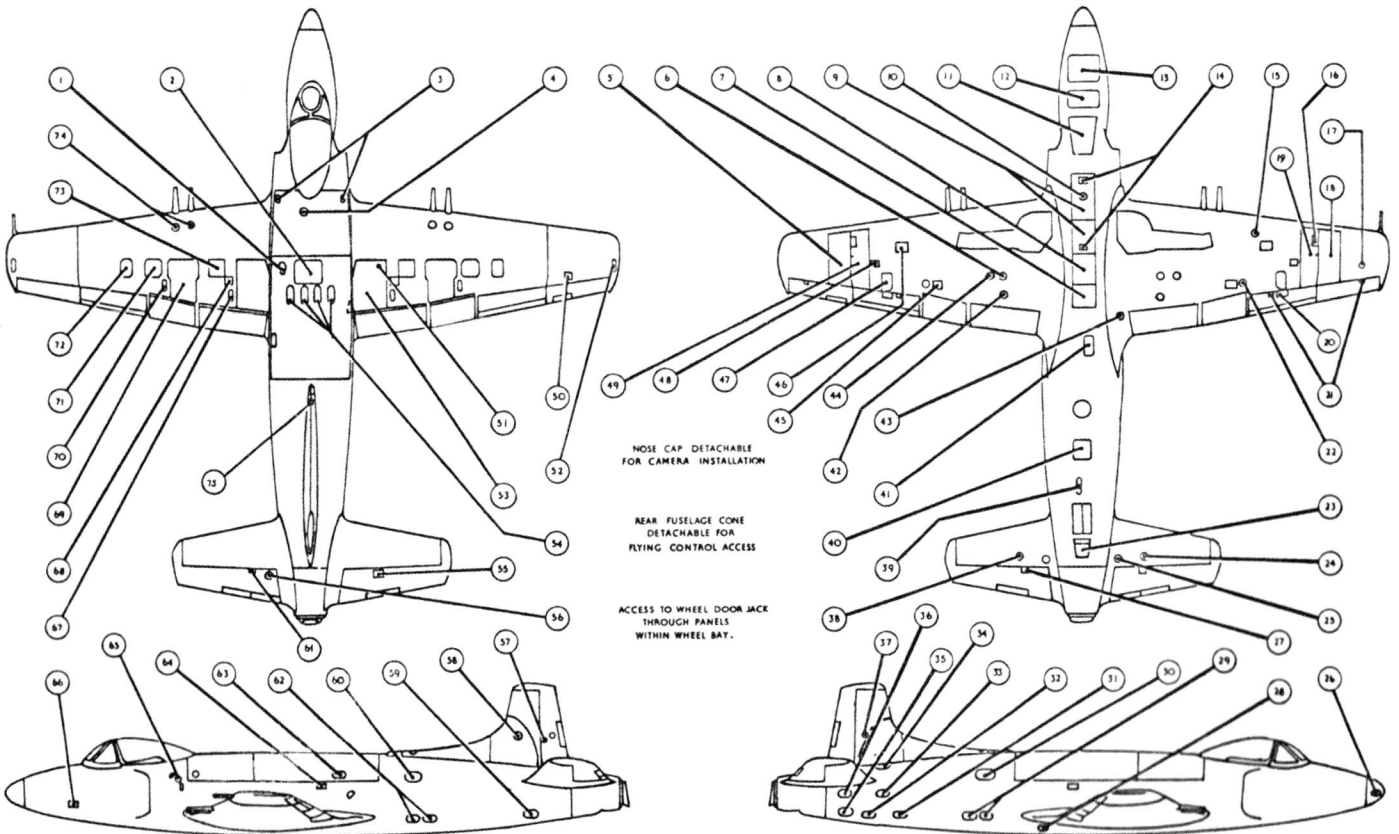

NOSE CAP DETACHABLE
FOR CAMERA INSTALLATION

REAR FUSELAGE CONE
DETACHABLE FOR
FLYING CONTROL ACCESS

ACCESS TO WHEEL DOOR JACK
THROUGH PANELS
WITHIN WHEEL BAY.

1 Rear fuel tanks filler
2 Access to accessory drive gearbox dip-stick, hydraulic header tank and engine oil filter
3 Forward side fuel tank fillers - port and starboard
4 Forward centre fuel tank filler
5 Wing fold mechanism - outboard portion
6 Hydraulic connections - port and starboard
7 Access to control trunk between Frames 13 and 14
8 Access to control trunk between Frames 12 and 13
9 Access to control trunk between Frames 8 and 11 and 11 and 12
10 Access to fuel drains
11 Access to forward control trunk
12 Access to forward control trunk
13 Access to forward hatch for radio and electrics
14 Drop tanks connections and fitting
15 Access to wheel illumination lamp - port only
16 Access to wing fold shear pin - port and starboard
17 Access to compass detector - port only
18 Wing fold mechanism - outboard portion
19 Wing fold mechanism - inboard portion
20 Access to aileron control rod - port and starboard

21 Aileron hinge locking and lubricating points - port and starboard
22 Electrical connections
23 Hold back gear
24 Access to elevator trim tab screw jack - port
25 Access to cables to elevator trim tab screw jack - port and starboard
26 Access to camera for reloading - starboard only
27 Access to elevator trim tab actuating rod - port and starboard
28 Lift spoiler jack
29 Access to jet tube connections, forward end and thermocouple connections - starboard
30 Access to jet turn and thermocouple connections - starboard
31 Tailwheel door jack auto sequence valve
32 Access to jet tube, aft end - starboard
33 Access to electrical connector bank - starboard
34 Access to electrical connector bank - starboard
35 Tailwheel door jack
36 Access to electrical connector bank - starboard
37 Access to 'Desynn' transmitter, rudder trim tab indicator - starboard
38 Access to elevator trim tab screw jack - starboard
39 Access to tailwheel hydraulics and castor-locking cable adjustment

40 Downwards identification lights
41 Access to control trunk, rear end, for flying control cable tensioning
42 Access to 'Desynn' transmitter flap indicator - port and starboard
43 Pitot line drain - port only
44 Access to flap interconnecting quadrants and cables - port and starboard
45 Aileron controls and flap jacks
46 Access to electrical connections and earthing points - port and starboard
47 Access to aileron control and pipe connections - port and starboard
48 Access to wing fold jack and electrical cables - port and starboard
49 Wing fold mechanism - inboard portion
50 Access to aileron trim tab actuator and 'Desynn' indicator outboard of fold - starboard only
51 Wing tank fuel filler
52 Access to wing tip lights - port and starboard and pitot head - port only
53 Wing tank bay cover - port and starboard
54 Ground running louvers
55 Access to 'Desynn' transmitter - elevator trim tab indicator - starboard only
56 Access to elevator spring tab mechanism - port only
57 Access to rudder trim tab actuating rod - port

58 Access to 'Miles' actuator, rudder trim tab - port
59 Access to jet tube connections, aft end and elevator cable tension compensators - port
60 Access to jet tube and thermocouple connections - port
61 Access to elevator trim tab actuating rod - port
62 Access to jet tube connections, forward end and thermocouple connections - port
63 Access to torch igniter - port only
64 RATOG electrical socket - port and starboard
65 Access to fuel pipe connections - port and starboard
66 Access to ground/flight switch, external electrical, pneumatic and oxygen charging
67 Access to flap jack - port and starboard
68 Access to fuel contents gauge amplifier - port and starboard
69 Gun door - port and starboard
70 Access to flap jack - port and starboard
71 Ammunition box - port and starboard
72 Ammunition box - port and starboard
73 Ammunition box - port and starboard
74 Gun hand holes - port and starboard
75 Jet pipe supporting screw

Sliding canopy.

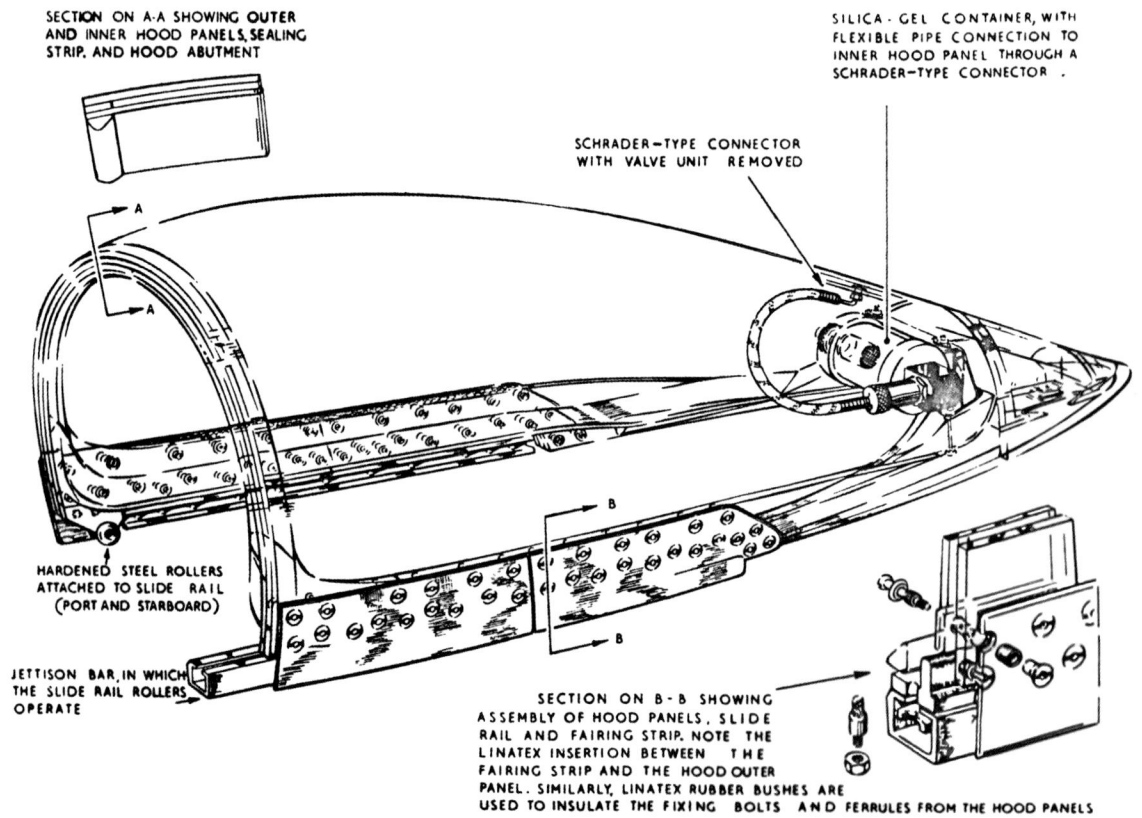

SECTION ON A-A SHOWING OUTER AND INNER HOOD PANELS, SEALING STRIP, AND HOOD ABUTMENT

SILICA-GEL CONTAINER, WITH FLEXIBLE PIPE CONNECTION TO INNER HOOD PANEL THROUGH A SCHRADER-TYPE CONNECTOR.

SCHRADER-TYPE CONNECTOR WITH VALVE UNIT REMOVED

HARDENED STEEL ROLLERS ATTACHED TO SLIDE RAIL (PORT AND STARBOARD)

JETTISON BAR, IN WHICH THE SLIDE RAIL ROLLERS OPERATE

SECTION ON B-B SHOWING ASSEMBLY OF HOOD PANELS, SLIDE RAIL AND FAIRING STRIP. NOTE THE LINATEX INSERTION BETWEEN THE FAIRING STRIP AND THE HOOD OUTER PANEL. SIMILARLY, LINATEX RUBBER BUSHES ARE USED TO INSULATE THE FIXING BOLTS AND FERRULES FROM THE HOOD PANELS

Windscreen.

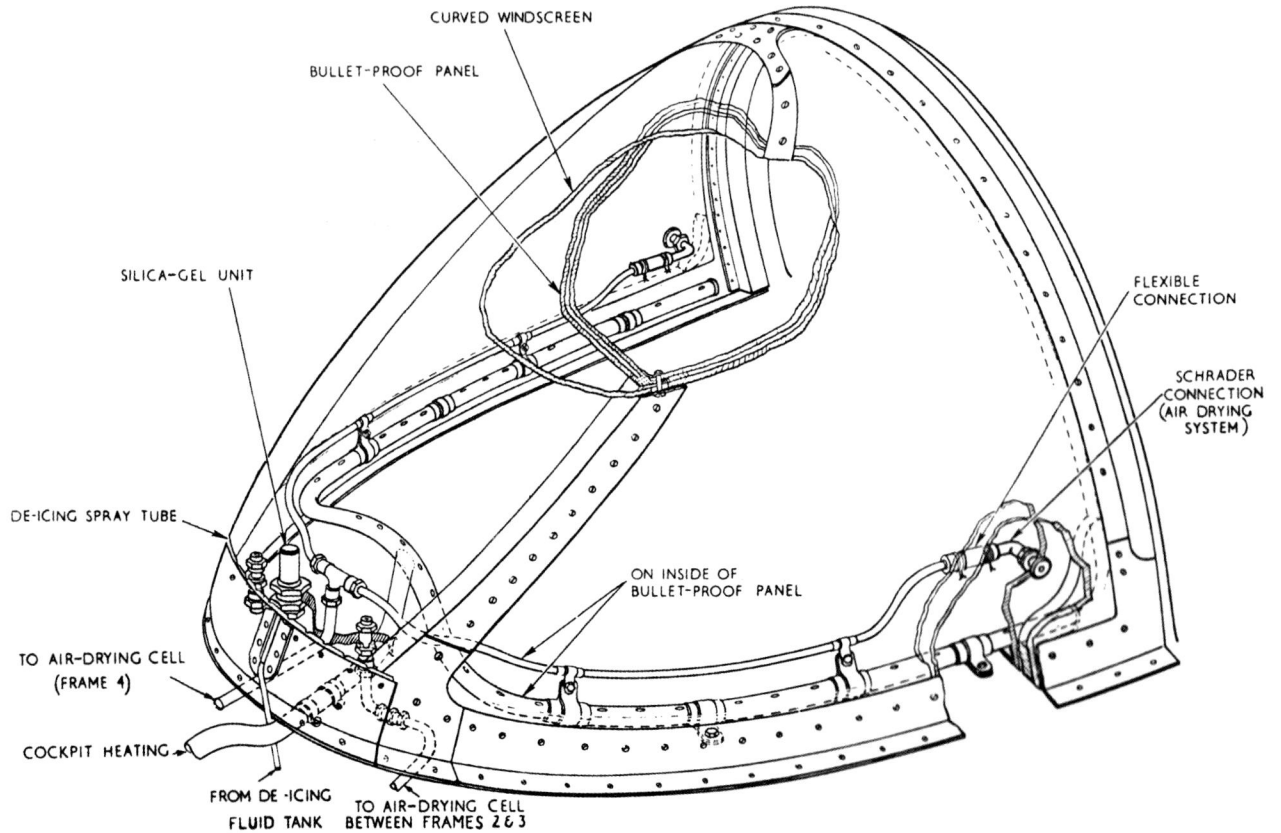

CURVED WINDSCREEN

BULLET-PROOF PANEL

SILICA-GEL UNIT

DE-ICING SPRAY TUBE

TO AIR-DRYING CELL (FRAME 4)

COCKPIT HEATING

FROM DE-ICING FLUID TANK

TO AIR-DRYING CELL BETWEEN FRAMES 2 & 3

ON INSIDE OF BULLET-PROOF PANEL

FLEXIBLE CONNECTION

SCHRADER CONNECTION (AIR DRYING SYSTEM)

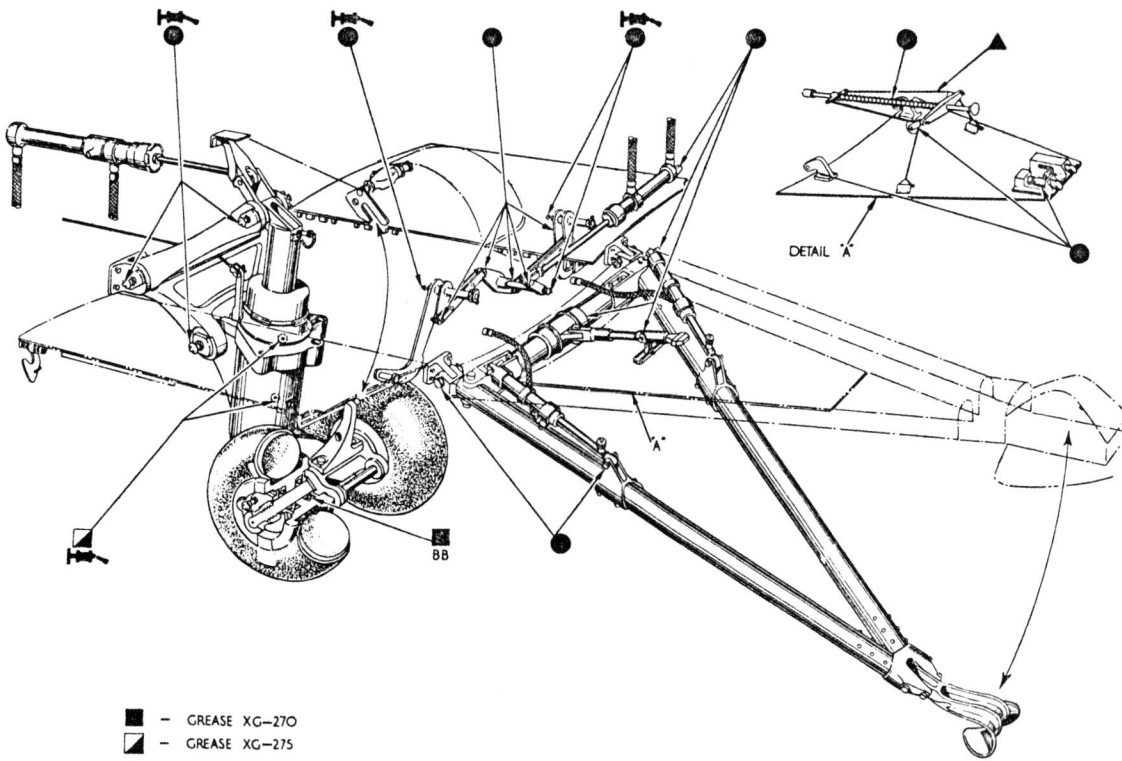

Tailwheel and arrester hook.

DETAIL 'A'

■ — GREASE XG—270
◪ — GREASE XG—275
▲ — WAX ZX
⟊— — GUN LUBRICATION
BB — BALL BEARINGS PACKED ON ASSEMBLY

Main wheel unit.

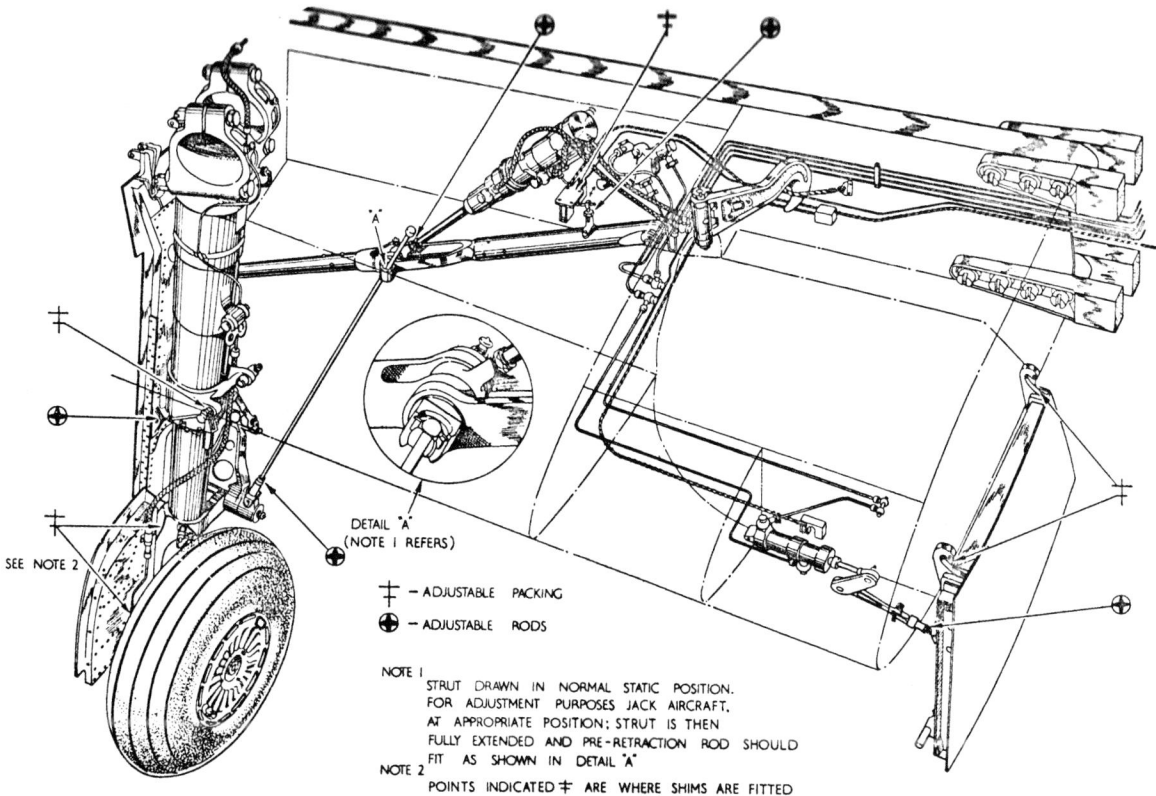

DETAIL 'A'
(NOTE 1 REFERS)

✛ — ADJUSTABLE PACKING
⊕ — ADJUSTABLE RODS

SEE NOTE 2

NOTE 1 — STRUT DRAWN IN NORMAL STATIC POSITION.
FOR ADJUSTMENT PURPOSES JACK AIRCRAFT.
AT APPROPRIATE POSITION; STRUT IS THEN
FULLY EXTENDED AND PRE-RETRACTION ROD SHOULD
FIT AS SHOWN IN DETAIL 'A'

NOTE 2 — POINTS INDICATED ✛ ARE WHERE SHIMS ARE FITTED

Rocket projectile installation.

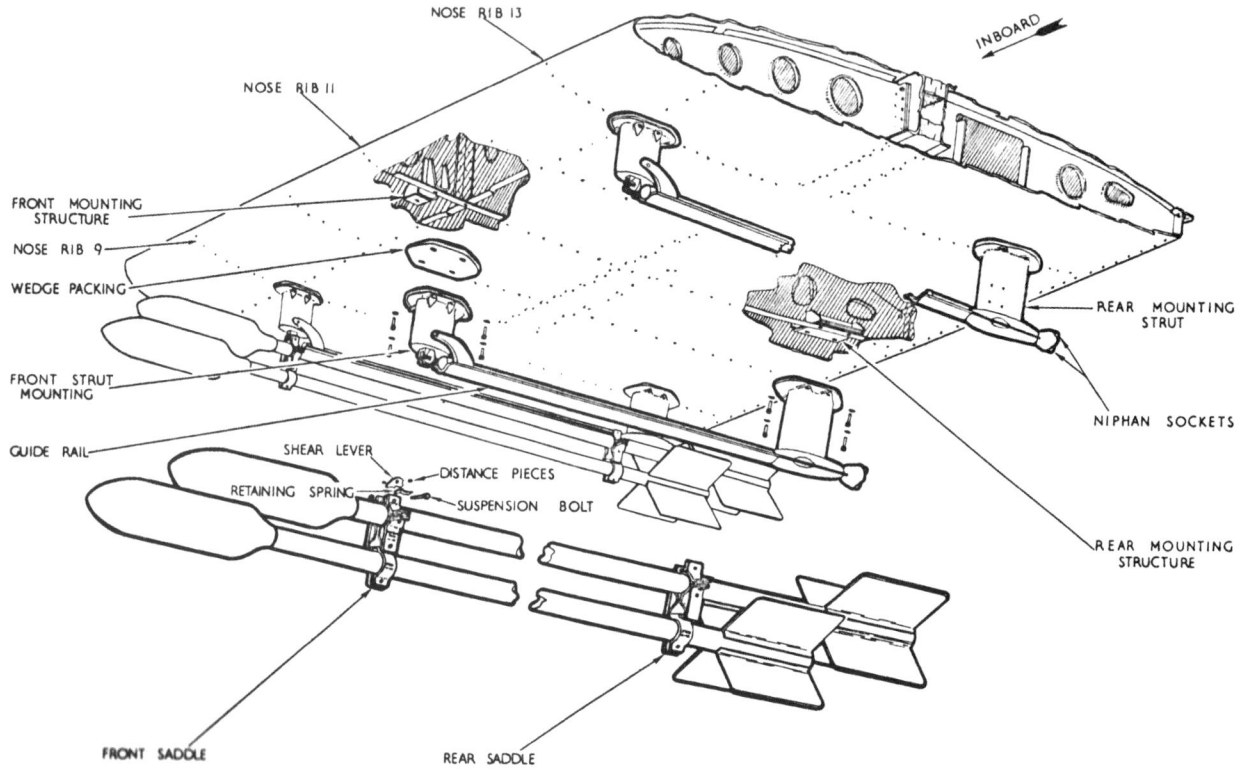

NOSE RIB 13
NOSE RIB 11
FRONT MOUNTING STRUCTURE
NOSE RIB 9
WEDGE PACKING
FRONT STRUT MOUNTING
GUIDE RAIL
SHEAR LEVER
RETAINING SPRING
DISTANCE PIECES
SUSPENSION BOLT
INBOARD
REAR MOUNTING STRUT
NIPHAN SOCKETS
REAR MOUNTING STRUCTURE
FRONT SADDLE
REAR SADDLE

Gun installation.

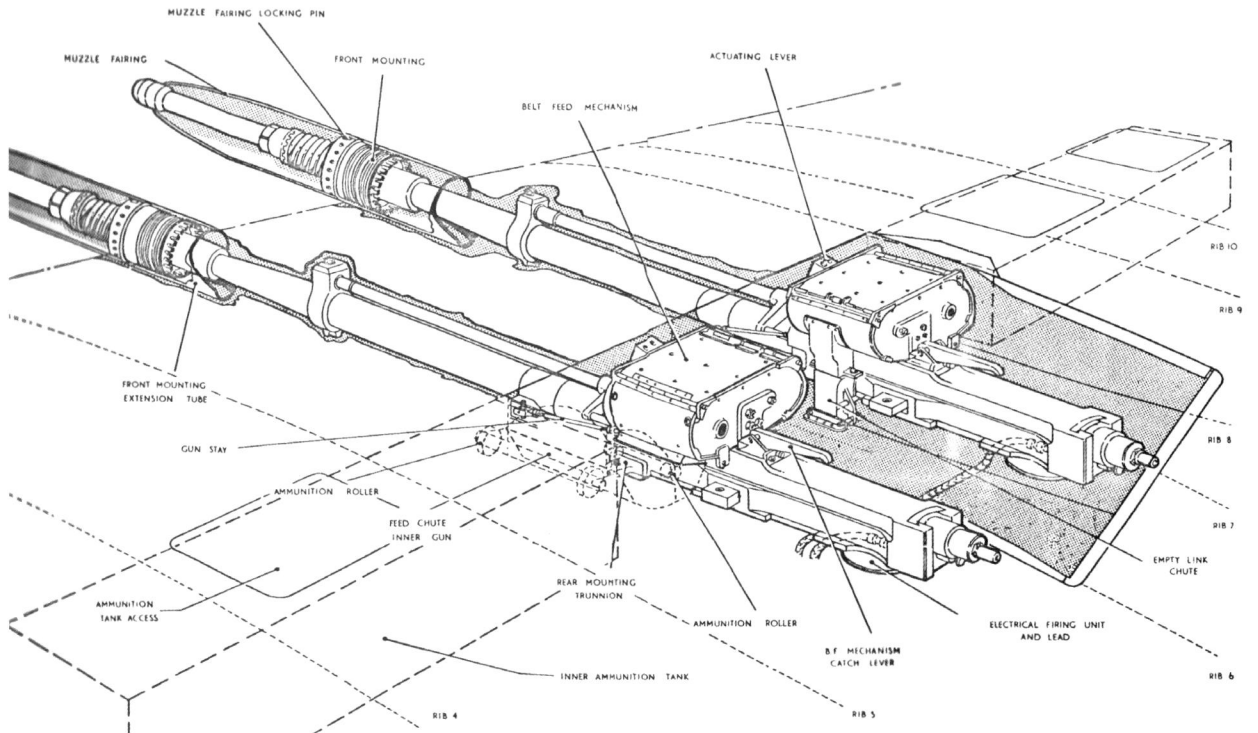

MUZZLE FAIRING LOCKING PIN
MUZZLE FAIRING
FRONT MOUNTING
BELT FEED MECHANISM
ACTUATING LEVER
FRONT MOUNTING EXTENSION TUBE
GUN STAY
AMMUNITION ROLLER
FEED CHUTE INNER GUN
AMMUNITION TANK ACCESS
REAR MOUNTING TRUNNION
AMMUNITION ROLLER
B.F. MECHANISM CATCH LEVER
INNER AMMUNITION TANK
RIB 4
RIB 5
RIB 6
RIB 7
RIB 8
RIB 9
RIB 10
EMPTY LINK CHUTE
ELECTRICAL FIRING UNIT AND LEAD

Rolls-Royce Nene Mk 3; later designated the 101 and 102.

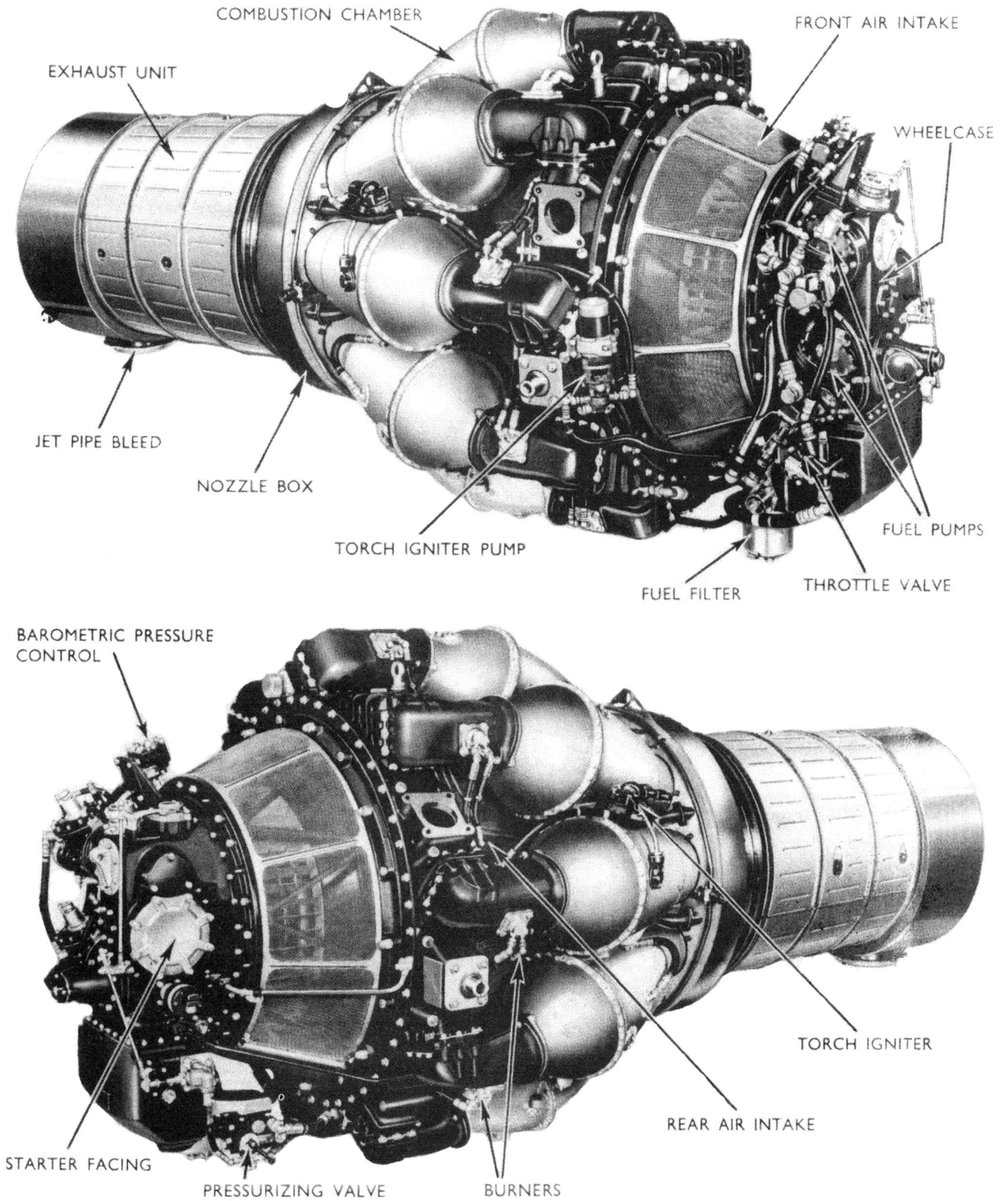

COMBUSTION CHAMBER

EXHAUST UNIT

FRONT AIR INTAKE

WHEELCASE

JET PIPE BLEED

NOZZLE BOX

TORCH IGNITER PUMP

FUEL FILTER

THROTTLE VALVE

FUEL PUMPS

BAROMETRIC PRESSURE CONTROL

TORCH IGNITER

REAR AIR INTAKE

STARTER FACING

PRESSURIZING VALVE

BURNERS

Appendix I: **TECHNICAL DATA**

We have refrained from using metric conversions in the following tables, as the Supermarine Attacker was built to Imperial specifications and dimensions and any conversion to metric would require working to many decimal places to be anywhere near accurate, therefore we have omitted any mention of metric. All references to fuel capacities are in imperial gallons.

Designation: F.1 / FB.1 / FB.2 / RPAF Attackers

Span:	36ft 11in, folded 28ft 11in (not RPAF)
Length:	37ft 1in (FB.2 and RPAF 37ft 6in)
Height:	9ft 11in
Engine:	One 5,000lb static thrust Rolls-Royce Nene 101 turbojet (FB.2 and RPAF 5,100lb st R-R Nene 102)
Fuel:	Fuselage centre tank 82
	Fuselage port side tank 73
	Fuselage starboard side tank 73
	Fuselage port rear tank 12
	Fuselage starboard rear tank 12
	Inner spar port wing tank 20½
	Inner spar starboard wing tank 20½
	Maximum internal fuel capacity 293
	External ventral tank 250
	(FB.2 and RPAF *internal* fuel capacity increased to 310)
	Maximum fuel capacity with ventral tank 543 (FB.2 and RPAF 560)
Weight:	Empty 8,434lb (FB.2 and RPAF 9,910lb)
Max speed:	590mph @ Sea Level. 538mph @ 30,000ft
Cruising:	380mph
Initial rate of climb:	
	6,350ft/min @ Sea Level

Ceiling:	48,500ft (FB.1 FB.2 and RPAF 45,000ft)
Range, nautical miles:	
	590nm without ventral tank
	1,190nm with ventral tank. (These figures are for F.1 only, range of later marks not to hand).
Radio & radar:	
	ARI (TR.1936, VHF), ARI.5307 (ZB/X beacon homing), A.I.1271 (beam approach), ARI.5679 (IFF) & ARI.5284 (radio altimeter)
Internal Armament:	All variants. Four 20mm Hispano Mk 5* cannon with 167 rounds (inboard) and 145 rounds (outboard)
External Armament:	
	FB.1 up to eight 60lb 3in rocket projectiles or four 300lb rocket projectiles or two 1,000lb bombs
	FB.2 and RPAF twelve 60lb 3in rocket projectiles or two 1,000lb bombs
Production:	**F.1**
	45. (63 F.1s ordered WA469-WA498 and WA505-WA537 however 16 were completed as FB.1s on the production line. Two F.1s cancelled WA536 and WA537).
	FB.1
	17. (11 WA525-WA535 + 5 WA481, WA507, WA510, WA511 and WA515 + 1 replacement WT851
	FB.2
	84. (WK319-WK342, WP275-WP304, WZ273-WZ302) + WA507 converted to FB.2
	RPAF
	36. R4000-R4035

F.1 WA493 106/J of either 800 or 890 Squadron the fuselage code remaining the same during its time with both squadrons. This aircraft was destroyed on 13th May 1952 when the pilot was forced to eject, the aircraft crashing at Binstead Park, Sussex, following an in-flight fire. FAA Museum

Above: *A photograph of an unidentified Attacker FB.2 believed to have been taken in 1954. This scene shows twelve rocket projectiles fitted beneath the wing all minus tailfins. Whether this is a trial installation or that the fins would be fitted later is not known.* via Ian Gazeley

Left: *This photograph again shows an FB.2 with a two tier rocket installation, presumably a 1,000lb bomb and rocket assisted take off gear.* via Tony Buttler

Appendix II: **ATTACKER SQUADRONS**

UK shore base location refers only to the period in which Attackers were operated by the respective squadron.

NO.700 SQUADRON. Refer to page 56.

NO.702 SQUADRON

First operated Attacker: RNAS Culdrose 03/52
UK shore base: Culdrose
Embarked onboard: N/A
Foreign bases: N/A
Disbanded: Culdrose 26/08/52, becoming 736 Squadron on the same date.
Reformed: Lee-on-Solent 30/09/57
Attacker variants operated: F.1/FB.1 03/52-08/52

NO.703 SQUADRON

First operated Attacker: RNAS Ford 07/51
UK shore base: Ford
Embarked onboard: HMS *Eagle* 18/04/55-21/04/55, HMS *Ark Royal* 27/06/55-2/07/55
Foreign bases: N/A
Disbanded: RNAS Ford 17/08/55
Reformed: 22/01/72 at RNAS Portland training Westland Wasp crews.
Attacker variants operated: F.1 07/51-08/51, FB.1 01/52-04/52, FB.2 04/55-17/08/55

NO.718 SQUADRON

First operated Attacker: RNAS Stretton 25/04/55
UK shore base: Stretton until 04/07/55.
Honiley 04/07/55-31/12/55
Embarked onboard: N/A
Foreign bases: N/A
Disbanded: Honiley 31/12/55
Reformed: N/A
Attacker variants operated: FB.2 03/52-08/52

NO.736 SQUADRON

First operated Attacker: RNAS Culdrose 08/52
UK shore base: Culdrose 26/08/52-04/11/53, Lossiemouth 04/11/53-08/54
Embarked onboard: N/A
Foreign bases: N/A
Re-equipped: Sea Hawk F.1 from 08/54
Disbanded: 26/03/65
Reformed: 03/65 with Buccaneer S.1
Attacker variants operated: F.1 08/52-08/54 FB.1 08/52-08/54 and FB.2 1953-08/54

NO.767 SQUADRON

First operated Attacker: RNAS Stretton 02/53
UK shore base: Stretton 02/53-03/55
Deck landing trials aboard: HMS *Triumph* 17-26/11/52, and 16-25/02/53, HMS *Ilustrious* 5-8/05/54
Foreign bases: N/A
Re-equipped: Sea Hawk F.1 from 02/54
Disbanded: Stretton 31/03/55
Reformed: 01/03/56 with Sea Hawk FB.3
Attacker variants operated: F.1 02/53-03/54, FB.1 02/53-??/53, FB.2 07/53-03/54

NO.787 SQUADRON

First operated Attacker: RAF West Raynham 01/51
UK shore base: West Raynham 01/51-09/54
Foreign bases: N/A
Disbanded: RAF West Raynham 16/01/56
Re-equipped: Supplemented by Sea Hawks of various marks remaining in service until squadron disbanded.
Reformed: N/A
Attacker variants operated: F.1 01/51-04/52, FB.1 09/52-09/54, FB.2 04/54-09/54

NO.800 SQUADRON

First operated Attacker: RNAS Ford, 21/08/51
UK shore base: Ford 21/08/51-04/03/52, 24/03/52-07/06/52, 19/07/52-04/09/52, 11/10/52-7/11/52, 3/12/52-26/01/53, 25/03/53-16/06/53, 16/07/53-02/09/53, 26/10/53-13/11/53, 30/11/53-03/02/54, 26/05/54-11/06/54
Embarked onboard: HMS *Eagle* 04/03/52-24/03/52, 07/06/52-19/07/52, 04/09/52-11/10/52, 7/11/52-3/12/52, 26/01/53-25/03/53, 16/06/53-16/07/53, 02/09/53-26/10/53, 13/11/53-30/11/53, 03/02/54-15/04/54, 27/04/54-26/05/54
Foreign bases: Hal Far, Malta 15/04/54-27/04/54
Disbanded: RNAS Ford, 06/54
Reformed: RNAS Brawdy 08/11/54 with Sea Hawk FB.3 08/11/54
Attacker variants operated: F.1, 08/51-05/52, FB.1 02/52-01/53, FB.2 09/52-06/54

NO.803 SQUADRON

First operated Attacker: RNAS Ford 26/11/51
UK shore base: Ford 26/11/51-04/06/52, 10/07/52-03/09/52, 09/10/52-07/11/52, 03/12/52-26/01/53, 25/03/53-16/06/53, 16/07/53-02/09/53, 23/11/53-03/02/54
Embarked onboard: HMS *Eagle* 04/06/52-10/07/52, 03/09/52-9/10/52, 07/11/52-03/12/52, 26/01/53-25/03/53, 16/06/53-16/07/53, 02/09/53-23/11/53, 03/02/54-14/04/54
Foreign Bases: Hal Far, Malta 14/04/54-16/06/54 and 26/06/54-13/08/54, Hyeres 13/08/54-17/08/54, Hal Far, Malta 19/08/54-10/54
Detachments: Idris, Egypt 16/06/54-26/06/54
Re-equipped: From 08/54 with Sea Hawk FB.3
Disbanded: 04/11/55
Reformed: 14/01/57 with Sea Hawk FGA.6
Attacker variants operated: F.1 11/51-01/53, FB.2 12/52-08/54 (last Attackers had left by 10/54)

NO.890 SQUADRON

First operated Attacker: RNAS Ford 21/04/52
UK shore base: Ford 21/04/52-27/10/52, Milltown 27/10/52-29/10/52
Embarked onboard: HMS *Eagle* 29/10/52-03/12/52
Foreign bases: N/A
Disbanded: HMS *Eagle*, 03/12/52
Reformed: RNAS Yeovilton on 20/3/54 with Sea Venom F(AW).20s
Attacker variants operated: F.1 04/52-10/52, FB.1 07/52-10/52, FB.2 07/52-12/52

Above: *Attackers of 800 and 803 Squadron ranged on the deck of HMS Eagle in the early 1950s. To their rear are Firebrands and early Fireflies with three bladed propellers. The two foremost Attackers are left, FB.1 WA531 and right F.1 WA496.* via Ian Gazeley

Below: *A poor quality but rare colour photograph taken from the squadron record book of one of the shortest lived Attacker squadrons, No.718. FB.2, WZ300, displays the squadron badge with numbers '7' and '8' visible either side of the torch, thus '718'. Variations of this squadron badge are believed to have included the letters 'V' and 'R' in lieu of the numbers '7' and '8' (see illustration on page 65). WZ300 later served with 1833 Squadron and as related previously, it overshot the runway at Ford and was subsequently broken up for spares.* FAA Museum

Opposite top: Three Attacker FB.2s from 803 Squadron seen performing a slow fly-by at Ford. It is possible that 143/J may be WK333. It is likely though that 150/J is WP303 and that 144/J is WK342. FAA Museum

Opposite bottom: An F.1 of 800 Squadron seen on the deck of HMS Eagle in rough seas off the coast of Norway in 1952. This machine is either WA487 or WA534, both of which were operated by the squadron during 1952. FAA Museum

NO.1831 SQUADRON RNVR

First operated Attacker: RNAS Stretton 14/05/55
UK shore base: Stretton 14/05/55-28/07/56, Brawdy 28/07/56-10/08/56, Stretton 10/08/56-10/03/57
Foreign bases: N/A
Detachments: Valkenburg 12/04/56-13/04/56
Disbanded: 10/03/57 as part of defence cuts
Re-equipped: N/A
Reformed: 03/04/80 as a training unit, but lapsed at later date.
Attacker variants operated: FB.2 14/05/55-10/03/57

NO.1832 SQUADRON RNVR

First operated Attacker: RAF Benson 08/55
UK shore base: Benson 08/55-11/56
Foreign bases: N/A
Detachments: Ford/HMS *Bulwark* 23/06/56-07/07/56
Re-equipped: Sea Hawk F.1 commencing early 1956.
Disbanded: 10/03/57 as part of defence cuts.
Reformed: 03/04/80 at RNAS Yeovilton as a training squadron, but lapsed at later date.
Attacker variants operated: FB.2 08/55-11/56

NO.1833 SQUADRON RNVR

First operated Attacker: RNAS Bramcote 10/55
UK shore base: Bramcote 10/55-23/10/55, Honiley 23/10/55-10/03/57
Foreign bases: N/A
Disbanded: 10/03/57 as part of defence cuts
Re-equipped: N/A
Reformed: N/A
Attacker variants operated: FB.2 10/55-03/57

OTHER UNITS KNOWN TO HAVE USED THE ATTACKER

No.700 Squadron, Trials and Requirements Unit
Ford, 17/08/55-15/02/56. Operated FB.2 WK277, possibly as a hack.

CIVILIAN - RN SUPPORT

Fleet Requirements Unit (FRU)
Hurn, 10/55-02/57. Operated FB.1 and FB.2.

Airwork Ltd
St.Davids, 12/55-07/57, Operated FB.1 and FB.2.

GOVERNMENT & RESEARCH

RAE Farnborough

TS409
Arrived at Farnborough by road from Vickers-Armstrong, South Marston on 26/09/51 for barrier trials there and on HMS *Eagle*. These were non-flying trials, as on all occasions TS409 was pulled into the barrier by cables. Allocated Instructional Airframe No.A2313, it left Farnborough by road for AHU Arbroath on 10/02/52.

TS416
Arrived at Farnborough on 27/03/50 for arrester trials, but returned to Vickers to have RATOG equipment fitted on 05/05/50. It returned to Farnborough on 10/07/50 to commence RATOG trials, returning once again to Vickers on 18/10/50 to have a steerable tailwheel installed, before flying back to Farnborough on 21/08/51 to continue with the RATOG trials. By 06/53 it was in use with the RAE for ground tests relating to deflected jet exhaust experiments, finally leaving Farnborough for RNAS St. Merryn for ground instructional use on 12/08/54.

A&AEE Boscombe Down
Various tests were carried out at this base during the career of the Attacker and the following aircraft are known to have visited or been tested there:

Prototype
TS409 –
TS413 Crashed fatally on 22nd June 1948.
TS416 –

F.1
WA469 Performance and handling trials.
WA471 Radio and rocket projectile clearance.
WA472 Unspecified tests.
WA485 Production flat-sided elevator trials, crashed fatally 5th February 1952.
WA535 Replacement for WA485 for flat-sided elevator trials.

FB.2
WK319 Mk 8 rocket projectile rail, bomb pylons and 5in HVAR rocket clearance.
WA525 Assisted WK319 in bomb pylon clearance.

ROYAL PAKISTANI AIR FORCE

The Attacker was only purchased by this one foreign nation. Little of the service history of these aircraft has come to light beyond that already discussed in chapter 4.

A mirror landing aid similar to those fitted aboard British aircraft carriers and designed to allow pilots to monitor their landing in relation to a ships deck, or in this case, at an airfield. FAA Museum

Appendix III: **COLOURS AND MARKINGS**

Throughout most of its service life the Attacker was painted in a standard colour scheme.

STANDARD COLOUR SCHEME

Upper surfaces:	Gloss Extra Dark Sea Grey
Undersides:	Gloss Sky
All lettering:	Black
Roundels:	Six positions

Type 392 & 398 (Specification. E.10/44) and Type 513 (Specification. E.1/45)

Initially all three prototypes, TS409, TS413 and TS416, were unpainted. Type C1 roundels with a thin yellow outline were applied to either side of the mid-fuselage, with Type C roundels above and below the wings. All three prototypes, while in natural metal finish, also carried the prototype 'P' in yellow within a yellow ring to the same diameter and just aft of the roundel on each fuselage side. Only the unpainted prototypes carried fin flashes, these being Type C.

In the early 1950s TS409 acquired a standard RN scheme of Extra Dark Sea Grey over Sky with the Type D roundels in six positions. The prototype 'P', in yellow, was retained on either side of the fuselage and the serial positioned under the tailplanes and each wing in black. When this aircraft entered the King's Cup Air Race at Hatfield in 1951 it carried the race number '5'. Under each wing the figure '5' was applied just inboard of the tip, at 90° to the serial with the top of the figure '5' orientated at the tip of the wing. The figure '5' was displayed in black on a white disc under both wings and white '5' on black for the fuselage and tail in lieu of national markings.

Squadron Service

In service the Attacker had a standard Royal Navy scheme of gloss finish Extra Dark Sea Grey (BS381C: 640) over Sky (BS381C: 210). Demarcation was high on the fuselage sides, with the entire dorsal spine, vertical fin and rudder in Sky. Serial numbers were applied in black 8in characters just under the leading edge of the tailplanes and repeated in 36in characters under the wings. Above the serial number on the rear fuselage, also in 8in black characters, was the legend 'ROYAL NAVY'. The underwing characters were orientated so that the serial under the starboard wing could be read from the front, and that under the port could be read from the rear. Three-digit unit code numbers were applied to either side of the fuselage, aft of the roundel on both sides. The style of the characters used varied from squadron to squadron, but most seem to be of a consistent 36in height. Unit

identification letters were sometimes applied to the top of either side of the vertical fin in 12in black letters. Examples of this are JA (later replaced with ST) for Stretton, CW (later replaced with CU) for Culdrose and LM for Lossiemouth. No photographic evidence seems to exist to prove that Ford applied the customary 'FD' codes to their aircraft, as many were detached aboard carriers and were allocated a single fleet letter code which represented the aircraft carrier instead of the shore base. These codes were individual identification letters, usually applied in 24in black characters on the vertical fin, e.g. J for HMS *Eagle*. Stencils were numerous and were usually in black characters on Sky areas and white characters on Extra Dark Sea Grey areas, although there were a few warning markings in red including the ejection seat warning triangle on either side of the forward fuselage below the canopy.

The undercarriage bays, oleo legs and inside of the undercarriage doors were all painted gloss Sky, as a means of protecting these areas from salt water. The wheel hubs were a dull aluminium.

Second-line and RNVR Units

Unique badges and emblems were not usual with front-line squadrons, but they came into vogue with second-line units and the RNVR squadrons. Probably the most well known is the winged greyhound emblem on either side of the nose of 1831 RNVR Squadron aircraft. This is often quoted as being grey, although most now seem to accept it was gold outlined in black. The CO of the unit had a slightly different version, as the head of the greyhound had a red 'bone dome' painted onto it! The exception to the rule as far as FAA squadrons go is 718 Sqn, which is known to have had the unit badge of a torch, with the letters 'V' and 'R' on either side of it, surmounted on a black bordered white diamond, applied to either side of the nose of their Attackers while at Stretton in 1955.

Squadron crests were often applied to Attackers, with the RNVR units usually carrying these above the station code letters on the vertical fin, while 787 Squadron for example had it applied to the forward fuselage, just ahead and below the windscreen on either side of the nose. 1833 RNVR Squadron had the noses of their aircraft painted red, the red tapering off to a point just in line and below the windscreen front edge, variations of which had appeared on 800 Squadron Attackers at an earlier date. A photograph also exists of an ex-736 Squadron machine from Lossiemouth being broken up at Bramcote in 1957 featuring a similarly painted nose. This does not extend as far aft as the version seen on 1833

Squadron aircraft and may have been applied in red or black. One or two of the first Attackers received by 1833 Squadron acquired 'BR' tail codes to go with fuselage codes in the '15x/16x' series. The 'BR' tail-codes do not appear to have been used when the '83x' codes were introduced.

The only non-service organisations to operate the Attacker were the A&AEE at Boscombe Down, RAE at Farnborough and Airwork Limited. All aircraft tested at Boscombe Down and Farnborough remained in their original markings, including the prototypes which were still unpainted. Those aircraft used by Airwork at Hurn and St.Davids or Brawdy within the Fleet Require-ments Unit were ex-squadron aircraft, therefore they retained the overall scheme and markings applied to Attackers at that time (1955). It is not known if all the aircraft supplied to them were from existing flying units or directly from storage or if either of these two sources refurbished them externally before Airwork received them. A study of period photographs seems to show that all of the aircraft they operated had the basic FAA camouflage and markings but their previous squadron badges and fleet/station codes had all been overpainted or removed. Six Attackers FB.2s were delivered to Hurn between 1st November 1955 (WZ286) and 18th January 1956 (WP285) and at the same time eight F.1s, FB.1s and FB.2s were delivered to Brawdy/St. Davids. In this period no codes seem to have been allocated, Airwork call-signs were used instead. After January 1956 however, in line with the new FAA code allocation system, Airwork were allo-cated codes 033 to 038 and for those at Brawdy/St. Davids BY/016 to BY/023. These numbers were applied in black in the same location as the three-digit service number had previously been applied, aft of the roundel. The characters used for this by the Brawdy/St. Davids aircraft seem to have been reduced in height from 36in possibly to 24in high, the unit code BY was applied in 12in black characters on the vertical fin. The three-digit characters on the Hurn-based aircraft seem to have been larger, although still not exactly the same size as the service-applied

codes. Images of Hurn-based aircraft confirm that no two-digit unit code was applied to these aircraft throughout their service. There is no evidence to date either to prove that FB.2, WP297, ever actually car-ried the code 035.*

The code allocation per airframe was as follows:

FRU Hurn	FRU Bawdy/St. Davids
WZ286 - 033	WA513 - BY/016
WK342 - 034	WA525 - BY/017
WP285 - 038	WA518 - BY/018
WP297 - 035*	WA519 - BY/019
WK337 - 036	WA531 - BY/020
WZ291 - 037	WA521 - BY/021
	WZ297 - BY/022
	WK333 - BY/023

Foreign Service

As already stated in chapter 4, the Attacker was only ever exported to Pakistan. All aircraft were prepared in the UK and finished in an overall gloss aluminium scheme, RPAF roundels were applied above and below the wings and either side of the rear fuselage. Pakistan's flag was applied to the base of the vertical fin, as per the location of the fin flash on RAF aircraft. Limited stencils, in English, can be seen on pho-tographs of the aircraft in the UK and after delivery, with the ejection seat triangle, sling and jacking points and camera access markings visible. The serial num-ber R4000 to R4035 was applied in black characters just aft of the fuselage roundel and these seem to be around 12in high. For flights in the UK British identifi-cation numbers were also temporarily applied. These were in the same colour and style as the aft serials and were positioned forward of the roundel e.g. G-15-209. Photographs of early aircraft (R4002 and R4003) transiting through Iran, show that these iden-tification numbers were no longer present. It is most likely therefore that these markings were only applied for test flights in the UK.

Left: *A poor quality but rare colour image of Attacker FB.2s of 1833 Squadron circa 1956. The fuselage roundels appear to be of different dimensions to those normally seen on Attackers* FAA Museum

Above: *F.1 WA513 seen here in storage awaiting disposal at Abbotsinch in 1958. This was an ex-FRU Brawdy/St.Davids machine, as can be seen by the 'BY' code on the vertical fin. Note the small dimensions and location of the three-digit aircraft code aft of the roundel. All Brawdy aircraft had the code applied in this manner e.g. WA531/020. via Author*

Below: *The standard scheme of the F.1 is well illustrated here. The 'J' tail code denotes HMS Eagle and the aircraft number (101) is representative of the size and style used, although there is no 'standard' for this marking. The aircraft is tethered by weights tied to a pole that passes through the 'sling' point in the aft fuselage. FAA Museum*

Opposite top: *This superb shot of 803 Squadron FB.2, WP292, '143' at RNAS Ford does show just how many stencils were on the basic airframe. You can see the panel under the nose, below the radio and battery area that is probably di-electric and thus unpainted. The anti-creep marks on the tyre/hub is of note, as is push-in button below the canopy and the lines hanging down from the wheel wells. FAA Museum*

Opposite bottom: *The effectiveness of the demarcation of the upper colour is well illustrated here with this mass of Attackers from Nos.800 and 803 on the deck of HMS Eagle. FAA Museum*

Seafang FR.32 VB895.
Flown by test pilot Mike Lithgow during deck landing trials on
HMS *Illustrious*, May 1947, and fitted with Rolls-Royce
Griffon 89 engine and contra-rotating propeller. VB895 was
struck off charge in August 1950 and her mortal remains are
believed to have perished at Shoeburyness in 1956. Extra
Dark Sea Grey/Sky finish with all lettering in black. 'C' type
roundels above wings

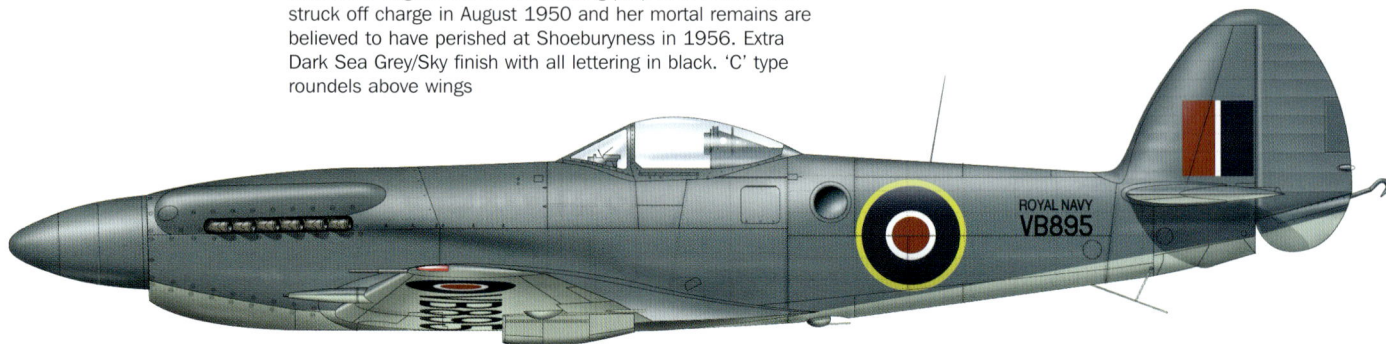

Supermarine Attacker Prototype TS409 post navalisation.
Natural metal finish with all lettering in black. 'C' type roundels in
six positions. Yellow 'P' within a yellow circle on both sides of rear
fuselage.

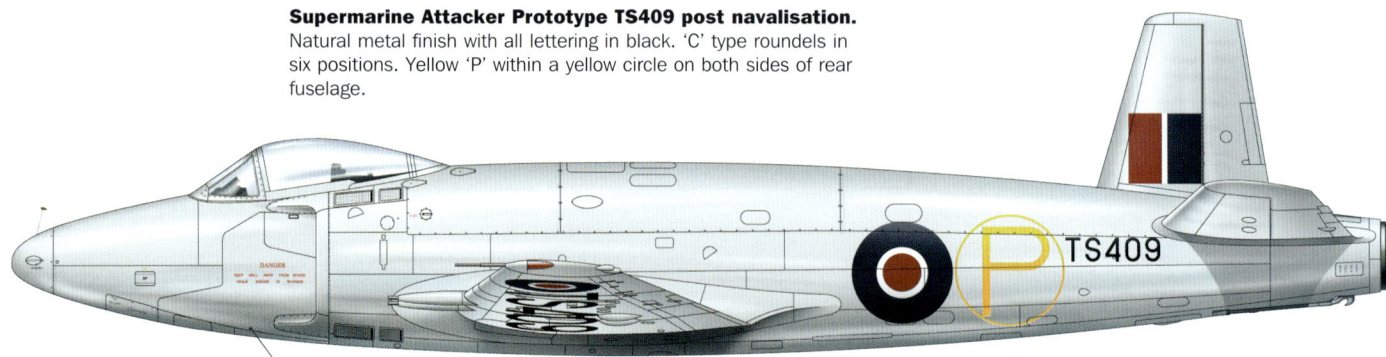

Supermarine Attacker Prototype TS409.
Bearing the race number '5' for the Kings Cup Air Race in
June 1951 at Hatfield. (The race was cancelled due to bad
weather). TS409 is minus weaponry but featuring a fin fillet.

Supermarine Attacker Prototype TS409.
Standard colour scheme. Yellow 'P' within a yellow circle on
both sides of rear fuselage.

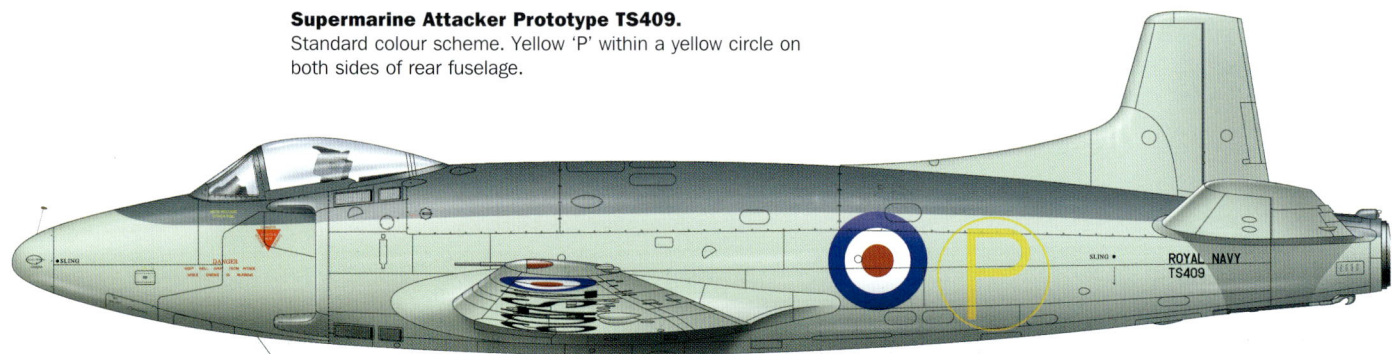

F.1

Supermarine Attacker F.1 WA469.
The first production aircraft delivered with original fin.
Standard colour scheme. On some earlier aircraft the Extra
Dark Sea Grey finished on the fuselage as shown.
See page 18.

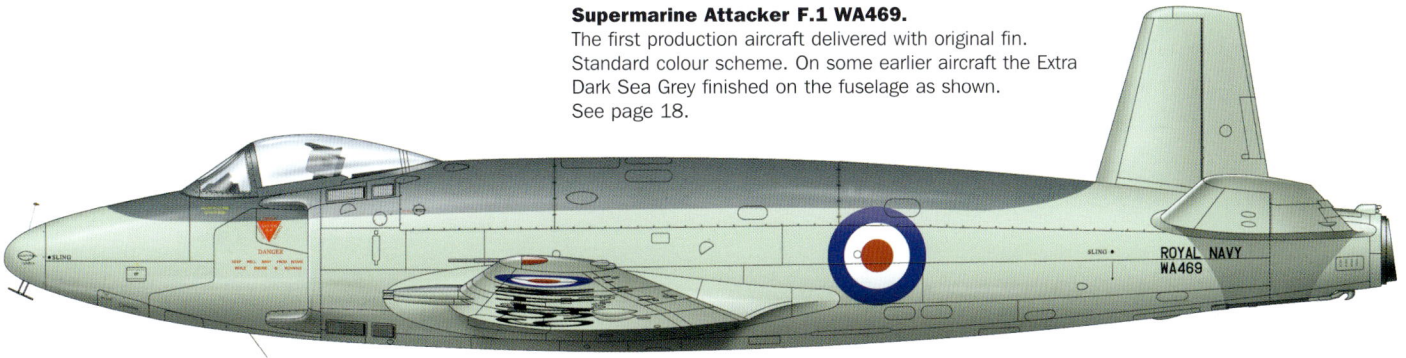

**Supermarine Attacker F.1 WA473, 102,
800 Squadron, August 1951.**
Standard colour scheme.

**Supermarine Attacker F.1 WA486, 105/CW,
736 Squadron, Culdrose, July 1952.**
Standard colour scheme. See page 22.

**Supermarine Attacker F.1 WA493, 106/J,
800 Squadron, HMS Eagle, 1952.**
This aircraft served with 800 Squadron between September
1951 and April 1952. Standard colour scheme.

FB.1

Supermarine Attacker FB.1 WA530, 150/LM, 736 Squadron, Lossiemouth, 1953.
This aircraft finished her days at Bramcote as Instructional Airframe A2363. Standard colour scheme. Red nose flash.

Supermarine Attacker FB.1 WA532, 106/J, 800 Squadron, HMS Eagle, early January 1953.
Standard colour scheme. Unit crest on fin, both sides.

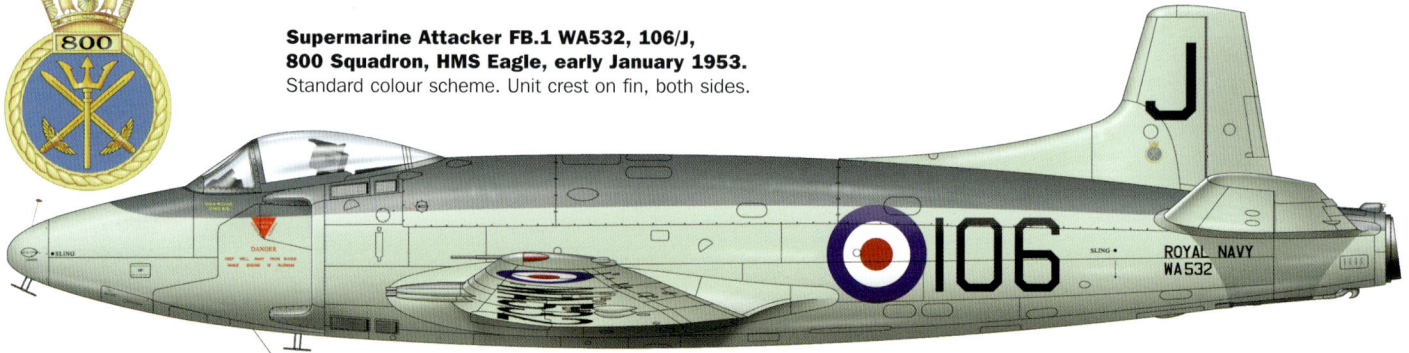

FB.2

Supermarine Attacker FB.2 WK320, 151/J, 890 Squadron, HMS Illustrious, 1952.
The second FB.2 built. Standard colour scheme. Unit crest below cockpit, both sides.

Supermarine Attacker FB.2 WP286, 101/J, 800 Squadron, HMS Eagle.
Standard colour scheme. Red nose flash. Unit crest on both sides of fin.

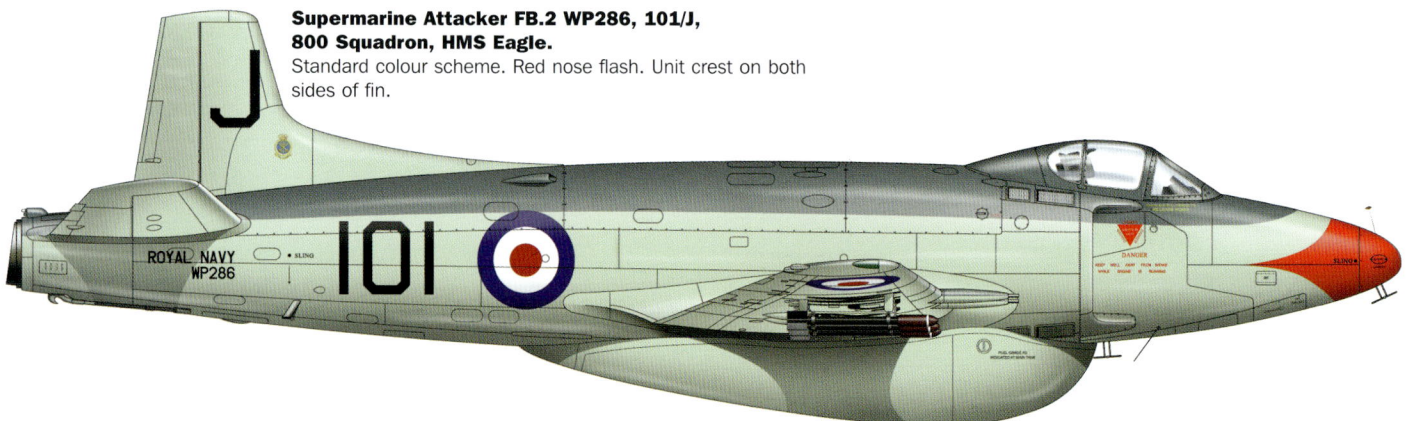

**Supermarine Attacker FB.2 WP303 150/J,
803 Squadron, HMS Eagle, 1953.**
Standard colour scheme. Unit crest below cockpit, both sides.

**Supermarine Attacker FB.2 WZ302, 163/ST,
718 Squadron, May/June 1955.**
Later transferred to 1833 Squadron, this maybe the one referred to in Don Chute's foreword regarding the only fatal accident which 1833 suffered whilst using the Attacker.

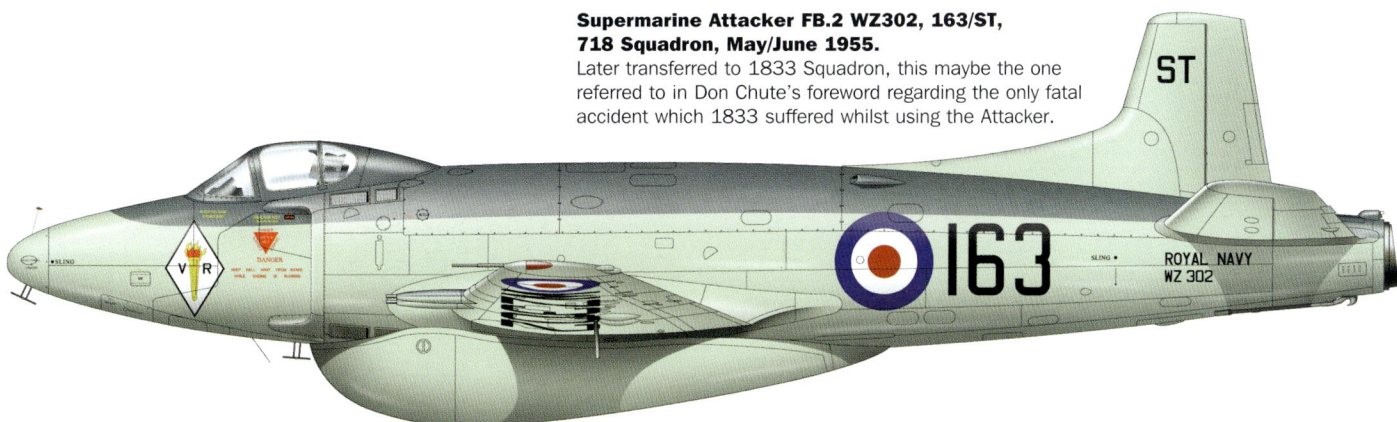

**Supermarine Attacker FB.2 WP289, 826,
1832 Squadron RNVR, 1956.**
Standard colour scheme. Figure '6' in black on front of belly tank.

**Supermarine Attacker FB.2 WP302, 836,
1833 Squadron RNVR.**
Standard colour scheme. Roundel Red flash on nose with unit crest superimposed. Note blanked off guns.

Supermarine Attacker FB.2 WZ294, 176/ST, 1831 Squadron RNVR, RNAS Stretton, 1955.
Standard colour scheme. 'Flying Greyhound' badge on nose in gold, outlined in black, both sides. Unit crest on fin. Note: A photograph exists showing WZ294 without the unit crest on the starboard side of the fin as shown on the starboard profile.

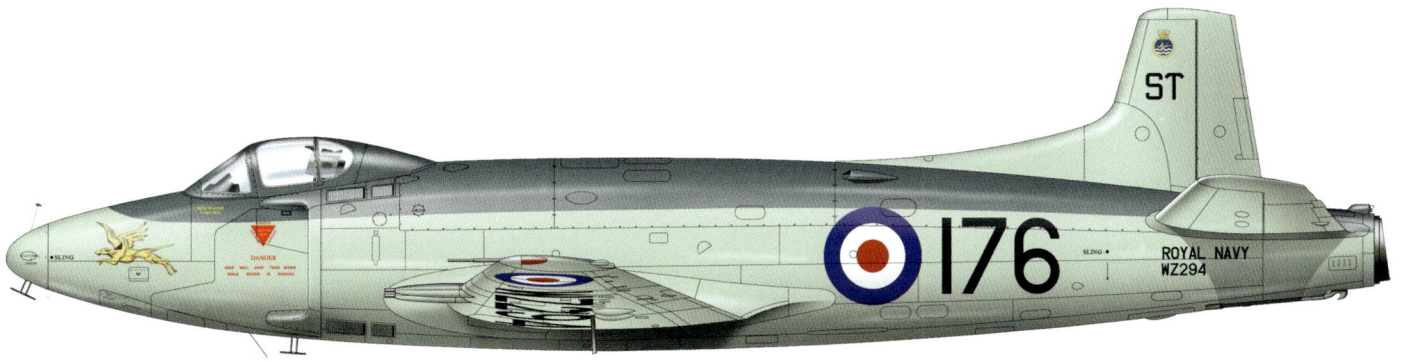

Supermarine Attacker FB.2 R4002.
One of 36 de-navalised examples delivered to the Royal Pakistan Air Force between June 1951 and May 1953. Aluminium overall with roundels in six positions, serial in black.

Supermarine Attacker F.1 / FB.1 and FB.2
1:72 Scale plans

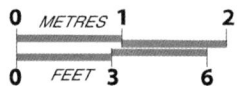

0 METRES 1 2

0 FEET 3 6

Supermarine Attacker F.1 / FB.1 / FB.2
Underside plan view.

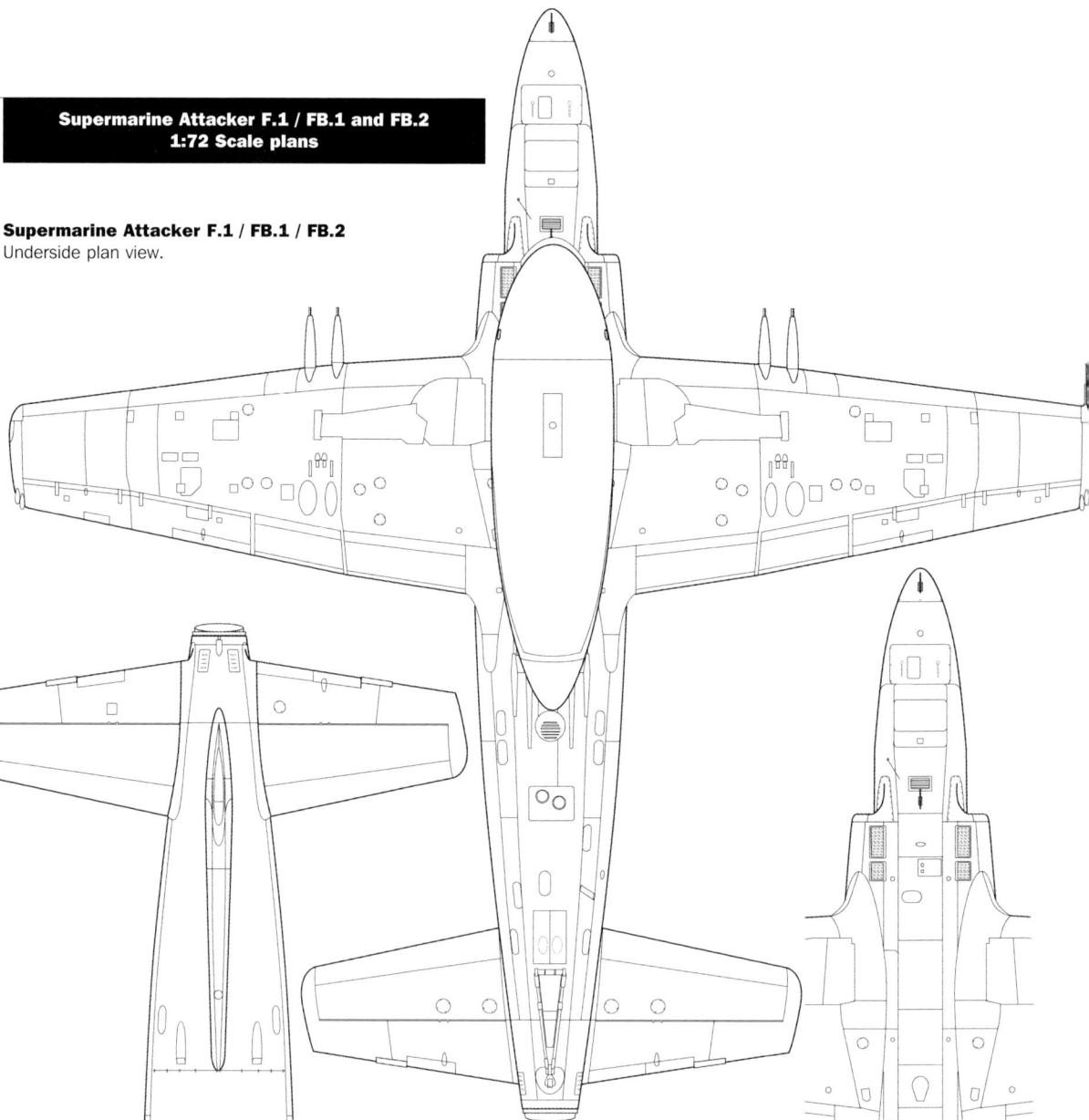

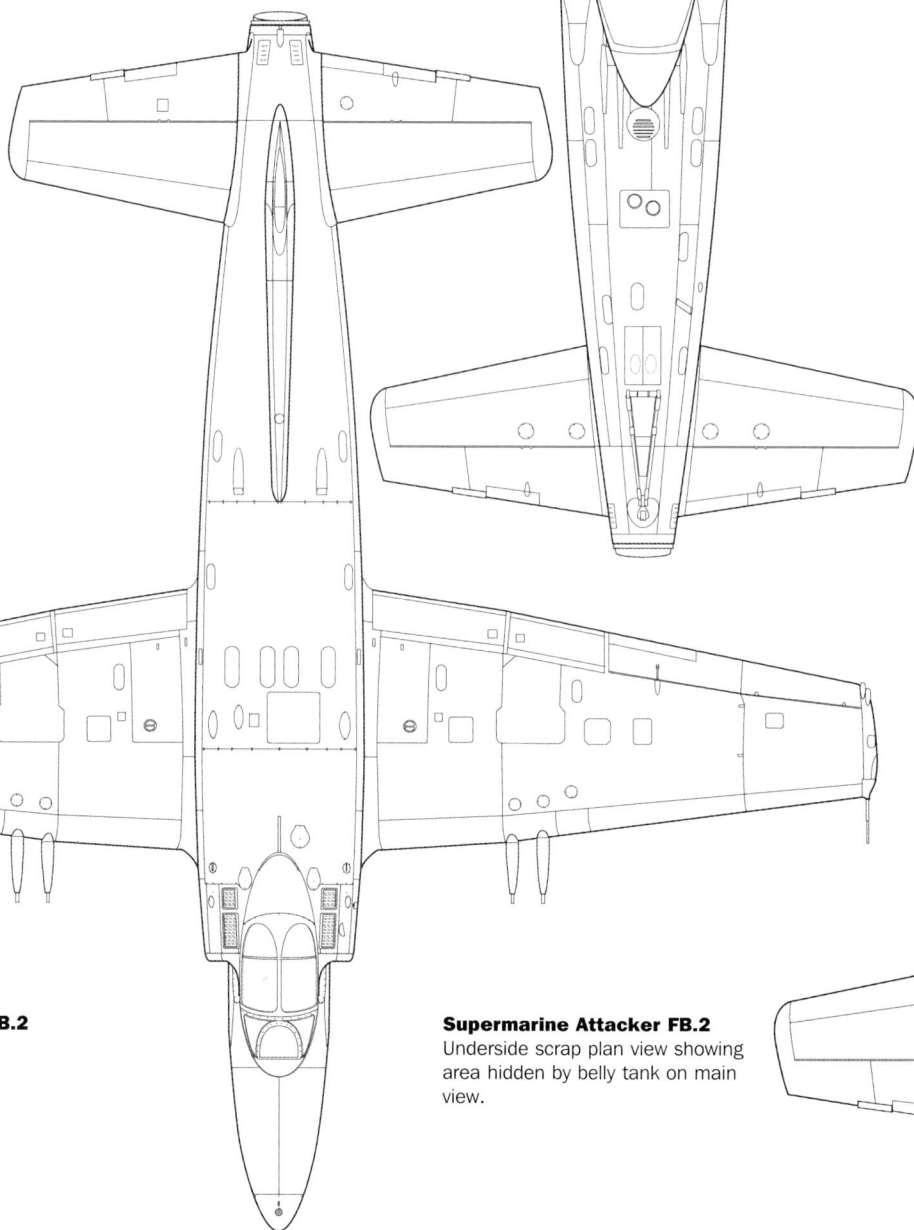

Supermarine Attacker FB.2
Upper plan view.

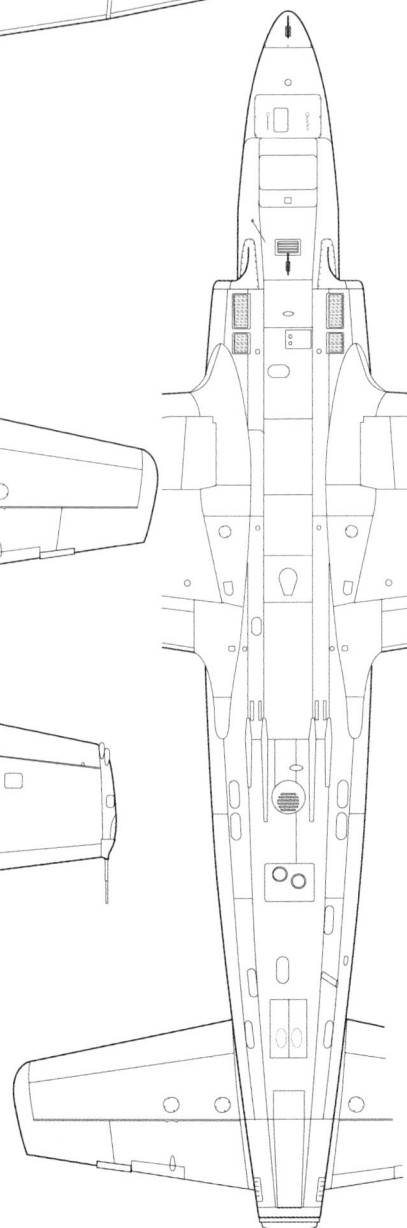

Supermarine Attacker FB.2
Underside scrap plan view showing area hidden by belly tank on main view.

**Supermarine Attacker
F.1 / FB.1**
Port profile. Shaded area
shows wing in folded position.

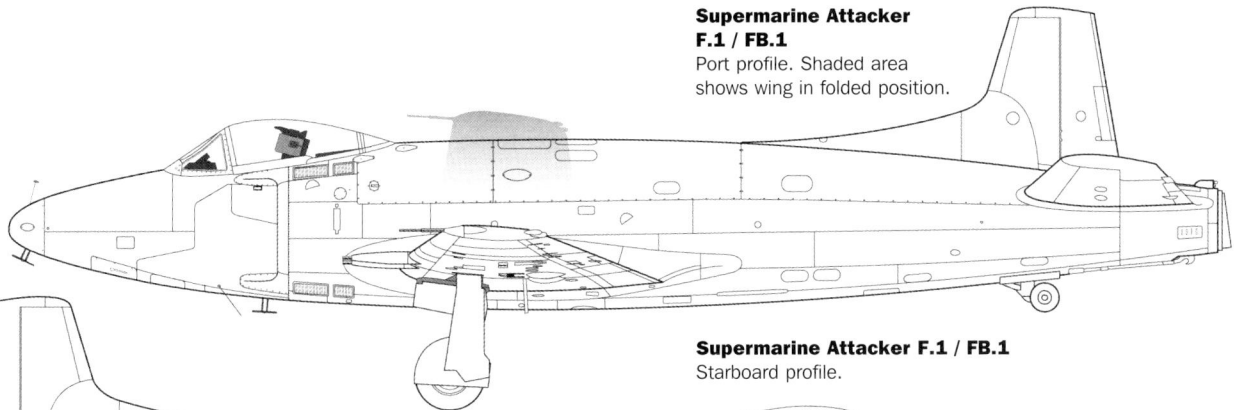

Supermarine Attacker F.1 / FB.1
Starboard profile.

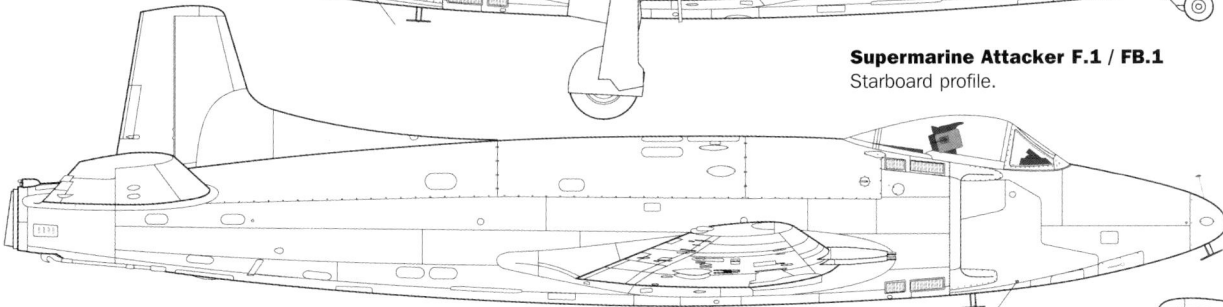

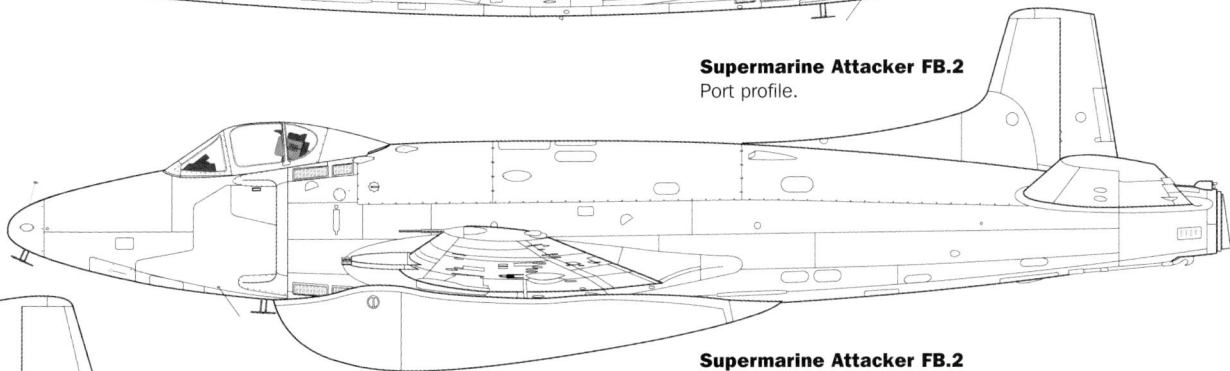

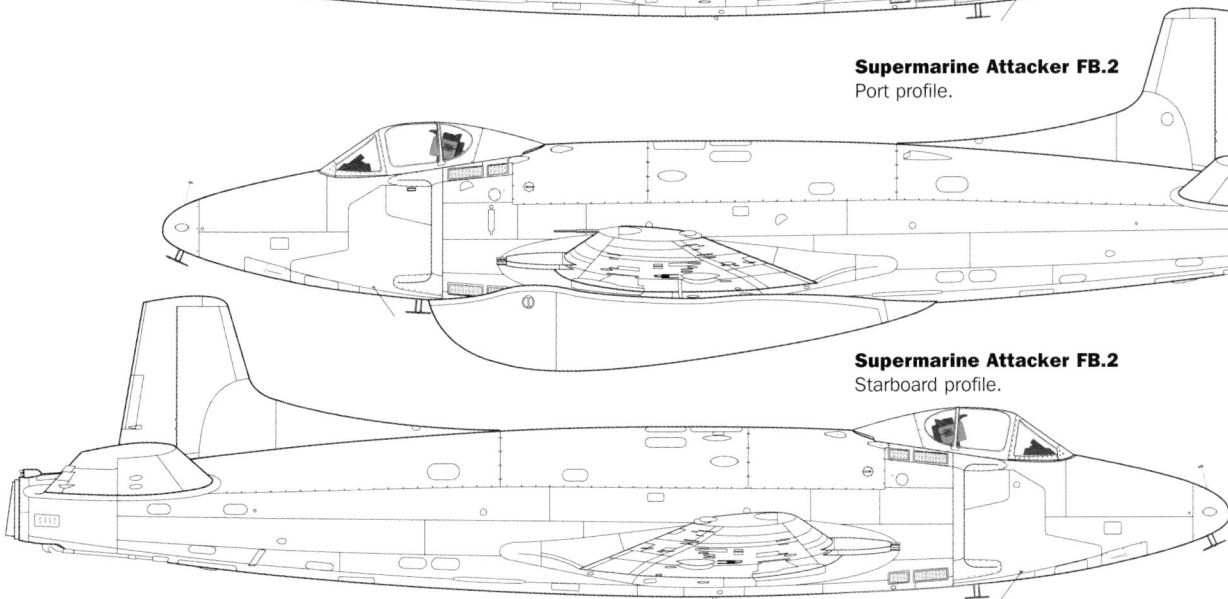

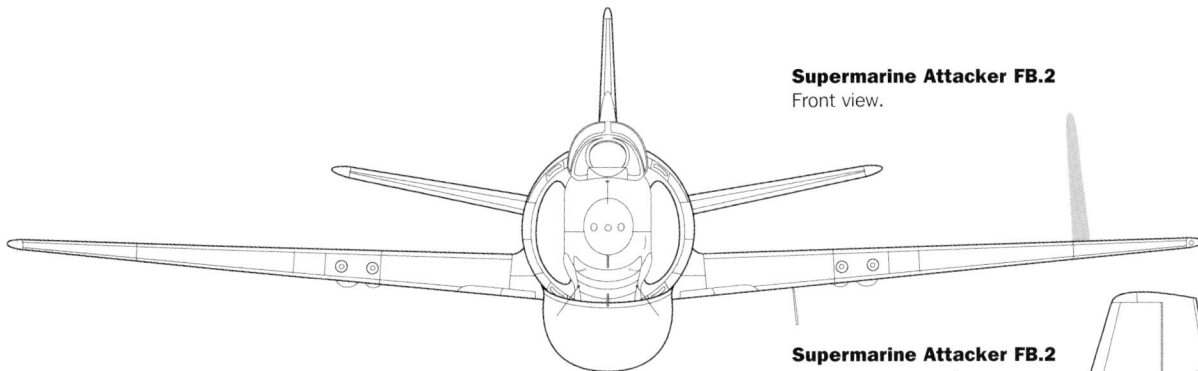

Supermarine Attacker FB.2
Port profile.

Supermarine Attacker FB.2
Starboard profile.

Supermarine Attacker FB.2
Front view.

Supermarine Attacker FB.2
Royal Pakistan Air Force.

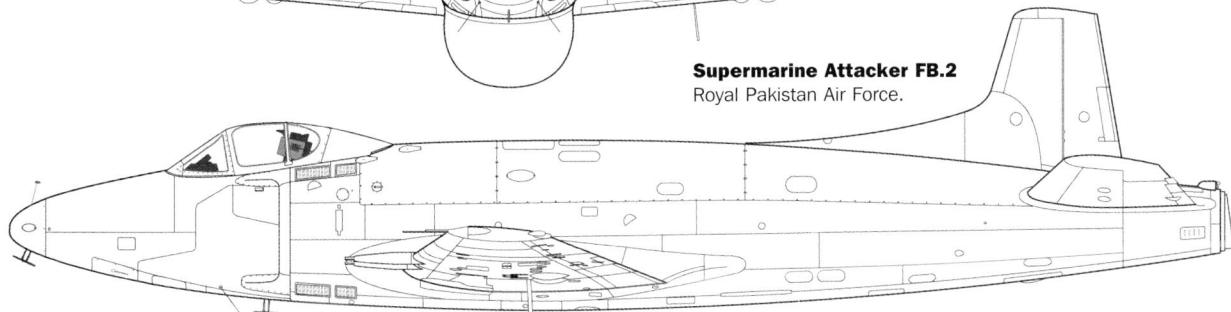

Appendix IV: **PRODUCTION**

Note: Bold serial denotes airframe extant.

Specification Number: E.10/44 & E.1/45
Contract Number: Air/4562/C.23(c)
E.10/44 Prototypes
Quantity: 3

TS409   Built to Specification. E.10/44 as Type 392 land-based version. Built at Hursley Park then by road to High Post. Taxying trials 17/06/46. By road to A&AEE Boscombe Down 1/07/46. FF 27/07/46. Engine failure, forced landing at A&AEE 26/08/46. Canopy lost during test flight 7/09/46. ASI failure, forced landing A&AEE 09/46. Appeared SBAC show, Radlett 09/46. Rtnd to Hursley Park for intake mods. Smaller fin and rudder fitted 04/47. To A Sqn, A&AEE for handling, stability and performance trials 09/06/47. To VA Chilbolton 23/06/47. Displayed at Lee-on-Solent 23/06/47. To A&AEE 30/06/47. To VA Chilbolton 3/07/47. Flew to Brussels/Melsbroek via Heathrow and Antwerp 4/07/47. Demonstration at Brussels then return to VA Chilbolton via Heathrow 06/07/47. To A&AEE 8/07/47. Aircraft once again in Belgium 3-6/08/47. Handling trials at A&AEE 09/47. To VA Chilbolton for mods 6/09/47. To SBAC Radlett 8/09/47. Rtnd to VA Chilbolton 15/09/47. To RAE Farnborough 11/10/47. Attained International 100km Closed-circuit Record at 564.88mph 25-27/02/48. VA Chilbolton to Bretigny via Hurn 18/04/48. Demonstrated to French Air Force then rtnd to VA Chilbolton via Hurn 20/04/48. Modified to Naval standards to replace TS413. FF in this form 5/03/49. To A&AEE 31/08/49. To RAE Farnborough by 3/10/49. To VA Chilbolton for repair 4/10/49. To C Squadron, A&AEE for dummy deck landing trials at Culdrose. Starboard u/c failed to lock, forced landing on grass at Culdrose, Cat. 4 damage 28/10/49. To VA Chilbolton for repair 28/10/49. To C Sqn, A&AEE for drop tank handling trials 6/01/50. Deck landing trials on HMS *Illustrious* 5/02/50. Tail leg collapsed during deck landing, Cat. 3 damage 8/02/50. VA Chilbolton for minor mods 17/02/50. To A&AEE for further drop tank trials 22/02/50. To VA Chilbolton for mods 28/04/50. Demonstration tour of the Middle East [Almaza, Cairo, Khlade, Beirut, Mezze, Damascus, Baghdad, Teheran, Ankara, Athens and Rome] 18/05/50-23/06/50. To A&AEE for spinning trials 5/07/50. National Air Races as No.46, won SBAC Challenge Cup 07/50. VA Chilbolton 28/07/50. Departed UK (Blackbushe) for Middle East tour 4/08/50. Electrical fault, Nicosia 5/08/50. Repaired on site by VA. Starboard u/c failed to lower on landing at Baghdad 7/08/50. Repaired on site by VA. To Sharjah 12/08/50. Lead aircraft in Pakistan Independence Day fly-past 08/50. Cap blew off fuel tank, landed at Almaza (Cairo) 4/10/50. Rtnd to VA Chilbolton 6/10/50. To A&AEE with new style elevator fitted 22/01/51. To VA Chilbolton for control mods 19/02/51. To A&AEE 6/03/51. Deck landings on HMS *Illustrious* 17-18/05/51. To VA Chilbolton 25/05/51. To have taken part in King's Cup Air Race as No.5, but cancelled due to bad weather at Hatfield 26/06/51. To A&AEE 2/07/51. Forced landing, undercarriage lever failure, Cat. 4 damage 2/08/51. To VA Chilbolton 3/08/51. RAE Farnborough for barrier trials on HMS *Eagle* [non-flying trial, ship in dock, a/c pulled through barrier by cables] 28/10/51. To RNAY Arbroath as GI Class II Instructional Airframe No.A2313 10/02/53. Broken up for scrap during 1956.

TS413   Built to Specification. E.10/44 as Type 398, first Naval prototype. FF VA Chilbolton 17/06/47. To C Sqn, A&AEE for handling & deck landing trials 27/08/47. RAE Farnborough 10/09/47. VA Chilbolton 12/09/47. RAE Farnborough 15/09/47. VA Chilbolton 16/09/47. Damaged during deliberate heavy landing at A&AEE, Cat AC 17/09/47. A&AEE 29/09/47. To VA Chilbolton 3/10/47. Deck landing trials from Ford 5/10/47. Deck landing trials on HMS *Illustrious* 14/10/47. Modification and makers trials. A&AEE Naval assessment trials 27/04/48. 703 Sqn, Thorney Island for deck landing trials on HMS *Implacable* 05-06/48. VA Chilbolton to A&AEE 2/06/48. VA Chilbolton fitted with drop tank 4/06/48. A&AEE 15/06/48. During measured take-off and handling trials a/c experienced violent yaw and dived into ground near Bulford village near Amesbury, Cat ZZ 22/06/48. Pilot (Lt T J A King-Joyce) was killed. Wreckage to VA Chilbolton 22/06/48.

TS416   Built to Specification. E.1/45 as Type 513 - improved Naval version. FF 24/01/50. VA Chilbolton to RAE Farnborough for

hook proofing 2/02/50. VA Chilbolton 3/02/50. Lee-on-Solent for Aerodrome Dummy Deck Landings 4/02/50. Tail leg attachment fractured 5/02/50. HMS *Illustrious* for deck landing trials (shore-based at Lee-on-Solent) 14/02/50. Port u/c fairing bent & wing skin damaged during landing on HMS *Illustrious* 16/02/50. VA Chilbolton 18/02/50. A Flt, RAE Farnborough for proofing of arrester gear 27/03/50. VA Chilbolton 5/05/50. RAE Farnborough 21/07/50. VA Chilbolton 18/10/50. A Flt, A&AEE for brake parachute, Mk 10 arrester gear and ground deflection of jet exhaust trials [non-flying]) 21/08/51. At RNAY St Merryn as GI Class II Instructional Airframe 21/05/54. On the fire dump at Culdrose by 07/58.

Specification Number: E.1/45
Contract Number: Acft/5530/C.23(c) dtd 9/07/45
Type 513 Naval version without wing fold
Quantity: 4

Initially allocated serial nos. VH987 to 990, but possibly latter amended to VH991 to 994 when the contract was issued on 21/11/45. Order cancelled.

Specification Number: E.1/45 (Naval)
Contract Number: Acft/5530/C.23(c) dtd 9/07/45
Type 513 Naval version with folding wings
Quantity: 14

Initially allocated serial nos.VH995 to 999 and VJ110 to 118, but possibly latter amended to 15 aircraft serial nos. VJ104 to 118 when the contract was issued on 21/11/45. Order cancelled.

Specification Number: E.1/45
Contract Number: 6/Acft/2822/CB.7(b) dtd 29/10/48
Type 511 F.1
Quantity: 63

WA469   Built VA South Marston. By road to VA Chilbolton 03/50. FF 5/04/50. Starboard wing tip folded during high speed trials at 200kts, emergency landing, port tyre burst, a/c swung off runway, Cat 4 (L R Colquhoun awarded George Medal) 23/05/50. C Flt, A&AEE for handling trials 1/08/50. VA Chilbolton 14/09/50. VA South Marston 5/11/50. C Flt, A&AEE for handling trials and gun firing 23/11/50. Canopy blew off during dive pull-out 1/05/51. Tested, VA Chilbolton 11/05/51. A&AEE 17/05/51. VA South Marston on CS(A) loan 29/05/51. C Flt, A&AEE for Mod 47 12/07/51. Engine failure in flight, relit, precautionary landing at A&AEE 30/07/51. VA Chilbolton for pressurisation tests 18/08/51. A&AEE 31/08/51. VA Chilbolton 26/10/51, VA South Marston to C Flt, A&AEE for gun firing tests at high altitude and Mach numbers 22/08/52. Rtnd VA South Marston for mods 5/11/52. C Flt, A&AEE 21/11/52. To ML Aviation, White Waltham 30/07/5. Moved by Northern Naval Aircraft & Salvage Unit to Ford 12/54. To RNAY Arbroath as GI Class II Instructional Airframe No.A2396 29/09/55. SOC 19/01/59.

WA470   Built to 'sub-standard' specification. A&AEE 11/50, Ford 5/11/50, HMS *Illustrious* for deck landing trials [based from Ford] 16/11/50. Trials with RN Test Unit, NAMDU Ford 14/12/50. VA South Marston 30/05/51. VA Chilbolton for mods 13/08/51. 703 Squadron, Ford for deck landing trials on HMS *Illustrious* 16-23/11/50. VA South Marston to 767 Squadron, Henstridge, swung on landing, went off runway, port u/c collapsed 11/01/52. VA South Marston deemed unrepairable 5/02/52. SOC, VA South Marston to GI Class II Instructional Airframe (No. N/K) 19/05/52.

WA471   Built to 'sub-standard' specification. RAE Farnborough for SBAC Display 4/09/50. VA Chilbolton 9/09/50. VA South Marston 10/09/50. W&E Flt, RAE Farnborough for radio clearance 4/10/50. VA South Marston for mods 8/12/50. C Flt, A&AEE for radio trials 12/01/51. VA South Marston via VA Chilbolton for mods and minor inspection 21/05/51. A Flt, A&AEE for handling and RATOG trials 12/10/51. Detachment, Ford for mock-up angled deck approach trials with HMS *Triumph* 10-13/02/52. VA South Marston 23/07/53. With Airwork General Trading Ltd at Gatwick for recondition by 11/54. AHU, Abbotsinch 21/03/57. Aircraft awaiting disposal/write-off 30/05/58. SOC 16/06/58.

WA472   Built to 'sub-standard' specification. VA South Marston to VA Chilbolton 18/10/50. C Sqn, A&AEE on temp loan for instruction of Service Trials Unit pilots 20/10/50. 703 Sqn, RNAS Ford 25/10/50. HMS *Illustrious* for deck landing trials

6/11/50. Hook bounced and tail cone damaged, HMS *Illustrious* 17/11/50. Rtn Ford, unserviceable 17/11/50. VA South Marston to Yeovilton 2/04/51. SOC as GI Class II Instructional Airframe 7/09/53.

WA473 Built to 'sub-standard' specification. Delivered RDU Culdrose 25/07/51. 800 Squadron (code 102/J), Ford 17/08/51: 702 Squadron (code 196/CW), Culdrose 20/04/52. Hydraulic accumulator burst, forced landing 14/05/52. 736 Squadron (code 196/CW), St. Merryn 26/08/52. VA South Marston 26/09/52. AHU, Abbotsinch 6/05/53. AHU, Culdrose 31/07/53. 736 Squadron (code 116/LM), Lossiemouth 22/09/53. Airwork General Trading Ltd, Gatwick, Cat.4 for recondition 30/08/54. With Airwork at Lasham by 09/56. AHU, Abbotsinch 12/12/56. Became gate guardian at Abbotsinch (code 146/J) 1957-1961 WOC 16/06/58. To FAA Museum RNAS Yeovilton 1961, Extant.

WA474 Built to 'sub-standard' specification. FF, VA Chilbolton 11/04/51. AWC 23/04/51. RDU, Stretton 24/04/51. Handling Sqn, RAF Manby for Pilot's Notes 8/05/51. 787 Sqn, West Raynham 3/07/51. Wheels-up landing, West Raynham 10/12/51. VA South Marston by road for repair 28/02/52. AHU, Abbotsinch 23/10/52. AHU, Culdrose 16/04/53. 736 Sqn, Culdrose 21/04/53. Hook broke, damage to stbd flap and fuselage by wire 22/04/53. 736 Sqn moved to Lossiemouth (code 113/LM) 9/11/53. Airwork Ltd, Gatwick, Cat.4 for recondition 20/07/54. AHU, Abbotsinch 19/04/56. To Eglinton by road for fire training 5/05/58. SOC 21/05/58.

WA475 Built to 'sub-standard' specification. FF 31/01/51. AWC 17/04/51. VA South Marston to RDU, Stretton 17/04/51. A Flt, RAE Farnborough on CS(A) charge for handling and RATOG trials 18/06/51. 702 Sqn, Culdrose 30/03/52. Engine failure on gliding approach, crashed on airfield, Cat.ZZ 4/06/52. SOC 28/07/52.

WA476 Built to 'sub-standard' specification. FF 20/11/50. Awaiting collection 8/01/51. Delivered to Handling Sqn, RAF Manby 9/01/51. Brakes failed on landing, u/c selected up to arrest a/c 22/01/51. VA Chilbolton 31/03/51. 702 Sqn, Culdrose 22/11/51. 736 Sqn 26/08/52. 702 Sqn, Culdrose 5/03/53. 736 Sqn, Culdrose (code 100/CW) 06/53. 736 Sqn, Lossiemouth (code 100/LM) 21/09/53. Airwork Ltd, Gatwick, Cat.4 for recondition 22/05/54. AHU, Abbotsinch 27/06/55. SOC (date N/K).

WA477 Built to 'sub-standard' specification. FF 3/02/51. 3rd production test a/c, went out of control during flight due to spoiler mechanism becoming damaged by a loose spanner, crashed 1 1/2 miles SW of Marlborough, P G Robarts, Vickers test pilot, killed 5/02/51-replaced by WT851 built to separate contract.

WA478 Built to 'sub-standard' specification. FF 7/01/51. AWC 22/01/51. To RDU, Stretton 22/01/51. 787 Sqn, West Raynham 6/02/51. Turbine blade failure, emergency landing at Greenham Common 03/05/51. VA South Marston by road for repair 8/05/51. 787 Sqn, West Raynham 31/08/51. Engine main bearing failure, forced landing Bircham Newton 1/12/51. 890 Sqn, Ford (code 142/J) 21/04/52. ARS, Culdrose 20/08/52. 736 Sqn, Culdrose 20/07/53. Airwork Ltd, Gatwick, Cat.4 for recondition but not proceeded with 16/10/53. To Gosport as Cat.5 (scrap) 10/03/55. SOC 14/03/55.

WA479 Built to 'sub-standard' specification. AWC 31/05/51. Delivered RDU, Stretton 11/06/51, 787 Sqn, West Raynham 27/06/51, 890 Sqn, Ford (code 148/FD & 148/J) 21/04/52. 736 Sqn, Culdrose (Code 106/CW & 106/CU) 15/08/52. 736 moved to Lossiemouth (code 106/LM) 6/11/53. Airwork Ltd, Gatwick by road, Cat.4 for recondition 31/05/54. AHU, Abbotsinch 20/03/56. To Eglinton for fire fighting training 5/05/58. SOC 21/05/58.

WA480 Built to 'sub-standard' specification. FF 7/01/51, AWC 31/01/51, Delivered RDU, Stretton 17/02/51, 787 Sqn, Ford, loud noise at 17,500ft, severe engine vibration, pilot ejected, a/c crashed Cat.ZZ 20/03/51. SOC 27/03/51.

WA481 Built to FB.1 specification. AWC 16/05/52, Delivered RDU, Culdrose 16/05/52, Gosport 10/06/52, Ford 18/06/52, 890 Sqn, Ford (code 145) 4/07/52, 800 Sqn, Ford (code 104/FD) 28/07/52, 890 Sqn (code 104) 15/09/52, 800 Sqn (code 104) 11/10/52, AHU Abbotsinch 15/01/53, Lossiemouth 25/03/54, 736 Sqn, Lossiemouth (code 100/LM) 3/05/54, AHU Abbotsinch for long term storage 13/07/54. To Bramcote by road as GI Class II Instructional Airframe No.A2362 19/11/54. SOC 14/10/58. Still present (code 100) 22/02/59.

WA482 Built to 'sub-standard' specification. FF 10/01/51, AWC 31/01/51, Delivered Handling Sqn, RAF Strubby 5/02/51, 787 Sqn, Ford overshot on 2nd approach during bad visibility 26/04/51. VA South Marston 9/05/51. RDU Culdrose 31/03/52, 890 Sqn, Ford (code 143/J) 12/05/52. AHU Culdrose 28/07/53-07/53. Airwork Ltd, Gatwick, Gat.4 for recon. 07/54. AHU Abbotsinch 20/10/54, Arbroath as GI Class II Inst. Airframe No.A2407 16/02/56. SOC 19/01/59.

WA483 Built to 'sub-standard' specification. FF 28/04/51, AWC 30/04/51, VA South Marston to RDU Stretton 3/05/51, 703 Sqn, Ford, heavy landing tail leg collapsed 2/07/51. To permanent CS(A) charge possibly for manufacturers tests 18/02/52. Airframe only from British Lion Production Assets Ltd to RAE Farnborough for barrier trials 27/12/52. Airframe only to Arbroath as GI Class II Instructional Airframe (possibly No.A2312) 10/02/53.

WA484 Built to 'sub-standard' specification. FF 27/04/51, AWC 31/05/51, Delivered RDU, Stretton 4/06/51, 703 Sqn, Ford (code 055/FD & 054/FD) 20/06/51. 703 Sqn to Belfast for steam catapult trials on HMS *Perseus* 28/07/51. Sqn rtnd to Ford 16/08/51. 800 Sqn, Ford (code 107/J) 23/08/51. 890 Sqn, Ford (code 147/J) 24/04/52. Lee-on-Solent for repair 27/05/52, 802 Sqn, Ford 28/08/52, VA South Marston for repair and inspection 15/10/52, AHU Abbotsinch 7/10/53, Lossiemouth 1/12/53. 736 Sqn, Lossiemouth (code 151/LM) 3/12/53. Airwork Ltd, Gatwick, Cat.4 for reconditioning 23/06/54, AHU Abbotsinch 1/02/56. Aircraft awaiting disposal/write off 20/05/58. SOC 16/06/58.

WA485 F.1 specification. FF 1/05/51. AWC 29/05/51, VA Chilbolton 29/05/51. Tests at VA Chilbolton abandoned 27/08/51. Further handling with flat-sided elevator tests 1-19/10/51. During rolls in circuit at 1,000ft aircraft entered sudden vertical dive and hit marshy ground near Leckford, 1 1/2miles west of VA Chilbolton, Cat.ZZ, pilot (L/C(E) R M Orr-Ewing) killed 05/02/52.

WA486 F.1 specification. AWC 31/05/51. Delivered RDU, Stretton 27/06/51. 787 Sqn, West Raynham 07/51, 890 Sqn, Ford (code 141/J) 22/04/52. 736 Sqn, Culdrose (code 105/CW & 105/CU) 23/07/52. 736 Sqn moved to Lossiemouth (code 105/LM) 6/11/52. Canopy shattered en-route to Airwork, aircraft landed safely at Abbotsinch 23/06/54. Airwork Ltd, Gatwick, Cat.4 for reconditioning 26/08/54. AHU Abbotsinch for long term storage 2/09/55. SFS 25/03/58.

WA487 F.1 specification. AWC 29/06/51, Delivered RDU, Culdrose 30/06/51, 800 Sqn, Ford (code 108/J) 31/08/51. During Deck Landing Practice on HMS *Eagle* tailwheel reversed, jetpipe and airframe damaged 12/03/52. Off-loaded to Ford 22/04/52. VA South Marston by road for repair 11/06/52, AHU, Abbotsinch 3/02/53. 767 Sqn, Stretton (code 173/J & 173/ST) 16/03/53. AHU, Lossiemouth 29/04/54. 736 Sqn. Lossiemouth 12/05/54. Airwork Ltd, Gatwick, Cat.4 for reconditioning 21/07/54. AHU, Abbotsinch for long term storage 7/02/56. SFS to Minworth Metals (Castle Bromwich) 30/05/58. WOC 16/06/58.

WA488 F.1 specification. AWC 31/05/51. Delivered RDU, Stretton 8/06/51. 787 Sqn, West Raynham 27/06/51. 800 Sqn, Ford (code 108, 106/J & 101) 7/09/51. 803 Sqn, Ford (code 117) 2/09/52. Tail oleo collapsed due to heavy landing, centre section and port wing buckled, HMS *Eagle* 19/09/53. VA South Marston by road for repair 29/09/52. AHU, Abbotsinch 4/06/53. AHU, Culdrose 14/07/53. 736 Sqn, Culdrose (code 104/CU) 25/07/53. 736 Sqn, Lossiemouth (code 104/LM) 9/11/53. Airwork Ltd, Gatwick, Cat.4 for reconditioning 30/06/54. AHU, Abbotsinch 5/03/56. SFS 30/05/58.

WA489 F.1 specification. AWC 29/06/51. RDU, Culdrose 2/07/51. 702 Sqn, Culdrose 1/05/52. 736 Sqn, Culdrose (code 112/CW & 113/CW) 26/08/52. With 736 Sqn to Lossiemouth (code 112/LM) 6/11/53. Airwork Ltd, Gatwick, Cat.4 for reconditioning 18/02/54. Dunsfold for acceptance checks 2/08/55. AHU, Abbotsinch for long term storage 17/08/55. SFS 30/05/58. WOC 13/06/58.

WA490 F.1 specification. AWC 27/06/51. Delivered RDU, Culdrose 29/06/51. 787 Sqn, West Raynham 07/51. VA South Marston for fuel tank change 10/51. 787 Sqn, West Raynham 16/10/51. Left formation at 35,000ft, anoxia suspected, crashed in sea 8 miles north of Brancaster, Cat.ZZ, pilot (L/C A.C. Lindsay) killed 06/03/52. SOC 17/03/52.

WA491 F.1 specification. AWC 30/06/51. Delivered RDU, Culdrose 4/07/51. 787 Sqn, West Raynham 16/07/51. VA Chilbolton on CS(A) charge 1/04/52. Airwork Ltd, Gatwick, Cat.4 for reconditioning 15/04/55. To Arbroath 14/12/56. To GI Class II Instructional Airframe No.A2423 16/01/57. SOC 19/01/59.

WA492 F.1 specification. AWC 30/06/51. Delivered RDU, Culdrose 3/07/51. 800 Sqn, Ford (code 104/J) 24/07/51. Overshot runway 7/11/51. 890 Sqn (code 144/J) 18/04/52. AHU, Culdrose 28/07/52. 737 Sqn, Culdrose 29/10/52. Aircraft went over the side on HMS *Eagle* when hook torn out and aircraft swung, pilot (Cdr P.F. Barker) killed. SOC 27/01/53.

WA493 F.1 specification. AWC 26/07/51. Delivered RDU, Culdrose 16/08/51. 800 Sqn (code 106/J) 26/07/51. 890 Sqn (code 106/J) 18/04/52. Fluid leak caused fire and controls to burn through, pilot (Lt T.F.B. Young) ordered to eject, a/c crashed Binstead Park, Sussex 13/05/52.

WA494   F.1 specification. AWC 25/07/51. Delivered RDU, Culdrose 25/07/51. 800 Sqn (code 105/J) by 09/51. To 890 Sqn (code 105/J) by 03/52. During landing practice tailwheel door struck by wire 04/03/52. Crashed in field near Bilsham Corner whilst landing at Ford, wings torn off, pilot OK 3/07/52. SOC 11/07/52.

WA495   F.1 specification. AWC 31/08/51. Delivered RDU, Culdrose 14/09/51. 703 Sqn, Ford 01/52. Bridle struck fuel tank during catapult launch from HMS Eagle 26/02/52. To 800 Sqn 04/52-07/52. AHU, Abbotsinch 01/53. 736 Sqn, Lossiemouth 9/03/54. Port wing tip folded during aerobatics, a/c crashed in sea nr Banff, Cat ZZ, pilot (Lt E.P. Tomlinson) killed.

WA496   F.1 specification. AWC 17/08/51. Delivered RDU, Culdrose 22/08/51. 800 Sqn, Ford (code 101/J) 31/08/51. AHU, Lee-on-Solent 29/05/52. 800 Sqn, Ford (code 117/J) 15/07/52. NAS Ford, stripped down and repainted 18/08/52. 803 Sqn, Ford (code 117) 11/10/52. AHU, Abbotsinch 29/01/53. Airwork Ltd, Gatwick by road, Cat.4 for reconditioning 8/02/54. AHU, Abbotsinch 21/10/54. To Bramcote by road 22/01/55. GI Class II Instructional Airframe No.A2373. SOC 14/10/58. Still in use 22/02/59.

WA497   F.1 specification. AWC 2/08/51. To C(A) charge at VA Chilbolton 2/08/51. Delivered to Rolls-Royce, Hucknall for noise suppression trials 29/08/51. VA South Marston 6/02/52. RDU, Culdrose 22/04/52. 890 Sqn, Ford (code 147/J) 12/05/52. 736 Sqn, Culdrose (code 114/CW) 25/08/52. Lost hook and went into barrier on HMS Eagle 14/09/52. ARS, Culdrose for repair 25/09/52. Ran off the runway after being caught in slipstream of Gloster Meteor, Culdrose 4/12/52. 736 Sqn, Culdrose (code 114/CW & 114/CU) 12/12/52. 736 Sqn, Lossiemouth (code 114/LM) 6/11/53. Airwork Ltd, Gatwick, Cat.4 for reconditioning 1/07/54. Hook lowered during landing, damaged Cat.L 2/07/54. Airwork, Dunsfold to AHU, Abbotsinch 20/01/56. SFS to Minworth Metals, Birmingham 16/06/58. SOC 6/06/58.

WA498   F.1 specification. AWC 22/08/51. Delivered RDU, Culdrose 23/08/51. 800 Sqn, Ford (code 103/J) 30/08/51. 800 Sqn, Ford 24/01/52. 890 Sqn, Ford (code 143/J) 18/04/52. 800 Sqn, HMS Eagle (code 103/J) 11/090/52. Hook pulled out, a/c hit barrier, wings torn off, pilot (L/C G.C. Baldwin) slightly hurt 21/09/52. Gosport for repairs 24/11/52 but not considerable viable and a/c scrapped.

WA505   F.1 specification. AWC 21/08/51. Delivered RDU, Culdrose 14/09/51. 803 Sqn, Ford (code 117/J) 15/11/51. Collided with hangar door, Ford workshops 22/08/52. AHU, Abbotsinch for long term storage 9/02/53. To Bramcote as GI Class II Instructional Airframe No.A2364 19/11/54. SOC 17/09/57.

WA506   F.1 specification. FF 25/08/51. AWC 30/08/51. Delivered RDU, Culdrose 11/09/51. 803 Sqn, Ford (code 1116/J) 29/11/51. AHU, Abbotsinch 4/02/53. Airwork Ltd, Gatwick by road, Cat.4 for repair 8/02/54. Airwork, Dunsfold to RDU, Stretton 21/02/55. To Arbroath 27/02/56. GI Class II Instructional Airframe No.A2408 7/03/56. SOC 19/01/59.

WA507   FB.1 specification. VA South Marston to CS(A) charge for Rolls-Royce, Hucknall for investigation into jetpipe rumble 28/04/52. VA South Marston converted to FB.2 25/03/53. AHU, Abbotsinch 4/06/53. Lee-on-Solent 11/01/54. NAMDU Gosport 18/01/54. AHU, Abbotsinch 31/05/54. 1832 RNVR Sqn, Benson (code 101 & 820) 22/08/55. AHU, Abbotsinch 14/09/56, SFS 25/03/58.

WA508   F.1 specification. AWC 31/08/51. Delivered RDU, Culdrose 17/09/51. VA Chilbolton for CS(A) trials 31/10/51. VA South Marston to AHU, Culdrose 15/11/51. VA South Marston 22/11/51. VA Chilbolton 9/11/53. Airwork Ltd, Gatwick, Cat.4 for reconditioning 20/04/54. Landed with hook down, Cat.SS 9/07/54. AHU, Abbotsinch 17/01/56. SOC 30/05/58. SFS 16/06/58.

WA509   F.1 specification. AWC 29/09/51, Delivered RDU, Culdrose 9/10/51. 703 Sqn, Ford (code 056/FD), power failure on approach, crashed in field short of runway, pilot (Lt J.C.E. Duncan) injured 23/01/52. SOC and used at Ford for instructional purposes 25/02/52.

WA510   FB.1 specification. AWC 21/05/52. Delivered RDU. Culdrose 21/05/52. 702 Sqn, Culdrose 9/07/52. 736 Sqn, Culdrose (code 103/CW, 103/CU & 103/LM) 26/08/52. Engine fire at 5,000ft, pilot (Sqn Ldr G.K. Gorton) ejected, a/c crashed 21m south of Port Gordon, Cat.ZZ 3/06/54.

WA511   FB.1 specification. AWC 22/05/52. Delivered RDU, Culdrose 22/05/52. 702 Sqn, Culdrose, fuel transfer valve problems, crashed on landing, pilot (Lt J.H. Nethersole) injured, Cat.ZZ 19/08/52. SOC 6/09/52.

WA512   F.1 specification. AWC 27/09/51. Delivered RDU, Culdrose 4/10/51. 803 Sqn, Ford (code 115/J) 27/11/51. Engine vibrations, precautionary landing 8/07/52. 890 Sqn, Ford 16/09/52. 803 Sqn (code 115/J) 22/10/52. AHU, Abbotsinch for long term storage 20/12/52. Removed from storage and test flown 2/04/54. Canopy flew off during loop,

Cat.L 7/05/54. Bramcote as GI Class II Instructional Airframe No.A2360 5/11/54. SOC 14/10/58. Still in use 22/02/59.

WA513   F.1 specification. AWC 29/09/51. Delivered RDU. Culdrose 4/10/51. 803 Sqn, (code 118/J) 27/11/51. Loaned to 890 Sqn (code 118/J) 15/09/52. 803 Sqn (code 118/J) 11/10/52. AHU, Abbotsinch by 01/53. 767 Sqn, Stretton (code 176/ST) 18/10/53-01/54. 736 Sqn, Lossiemouth (code 176/LM) 03/54-07/54. AHU, Abbotsinch by 03/55. Airwork Ltd, Gatwick, Cat.4 for reconditioning by 05/55. Dunsfold to AHU, Abbotsinch 8/07/55. Airwork, St.Davids (code 016/BY) 8/12/55. AHU, Abbotsinch 1/07/57. SFS 25/03/58

WA514   F.1 specification. AWC 30/09/51. Delivered RDU, Culdrose 9/10/51. 803 Sqn, Ford (code 113/J) 21/11/51. AHU, Abbotsinch 16/01/53. Airwork Ltd, Gatwick by road, Cat.4 for reconditioning 8/02/54. AHU, Abbotsinch 3/02/55. SFS 25/03/58.

WA515   FB.1 specification. AWC 20/05/52. Delivered RDU, Culdrose 21/05/52. 702 Sqn, Culdrose (code 197) 18/07/52. 736 Sqn, Culdrose (code 107/CW & 107/CU) 26/08/52. 736, Lossiemouth (code 107/LM) 9/11/53. Airwork Ltd, Gatwick, Cat.4 for reconditioning 27/08/54. AHU Abbotsinch for long term storage 9/11/53. SFS 16/06/58.

WA516   F.1 specification. AWC 30/09/51. Delivered RDU, Culdrose 9/10/51. 803 Sqn, Ford (code 112/J) 27/11/51. Aircraft overshot runway into field after failing to get airborne, badly damaged 14/10/52. ARS, Ford 14/10/52. AHU, Abbotsinch 24/02/53. Airwork Ltd, Gatwick by road, Cat.4 for reconditioning 8/02/54. AHU, Abbotsinch for long term storage 8/02/55. Bramcote in use as Instructional Airframe (No. N/K) by 11/58.

WA517   F.1 specification. AWC 24/10/51. Delivered RDU, Culdrose 30/10/51. VA South Marston 11/06/52. 736 Sqn, Culdrose 29/08/52. No.3 aircraft in stream take-off, caught in slipstream of preceding a/c. flicked to port and hit runway, caught fire, pilot (Lt C J M Barclay) killed. SOC 19/03/53.

WA518   F.1 specification. AWC 25/10/51. Delivered RDU, Culdrose 1/11/51. 803 Sqn, Ford (code 114/J) 27/11/51. 890 Sqn, Ford (code 114/J) 11/09/52. 803 Sqn, Ford (code 114/J) 22/10/51. AHU, Abbotsinch for long term storage 20/12/52. Airwork Ltd, Gatwick by road, Cat.4 for reconditioning 8/03/54. Airwork, Dunsfold to AHU, Abbotsinch 15/03/55. Airwork, St.Davids (code 018/BY) 29/11/55. AHU, Abbotsinch for long term storage 7/03/57. SOC 23/03/58. At Bramcote as Instructional Airframe (No. N/K) 11/58.

WA519   F.1 specification. FF 28/10/51. AWC 31/10/51. Delivered RDU, Culdrose 6/11/51. 803 Sqn, Ford (code 111/J) 27/11/51. AHU, Abbotsinch 20/12/52. 767 Sqn, Stretton 26/10/53. Airwork Ltd, Gatwick, Cat.4 for reconditioning 21/12/53. AHU, Abbotsinch 6/05/55. Airwork, St.Davids (code 019/BY) 18/10/55. Swung off perritrack and hit marshalling box, Cat.LX 7/05/56. AHU, Abbotsinch for long term storage 22/02/57. SFS 25/03/58.

WA520   F.1 specification. AWC 31/10/51. Delivered RDU, Culdrose 7/11/51. Long term storage 12/12/51. 736 Sqn, Culdrose 19/01/53. Forced landing on airfield after pilot failed to select internal tank 4/03/53. ARS, Culdrose 4/03/53. VA South Marston by road 21/043/53. Airwork, Gatwick by road 14/04/53. AHU, Abbotsinch 18/03/54. Arbroath as GI Class II Instructional Airframe No.A2402. Remains were with Eyres & Co., Mitcham, Surrey by 15/04/58. SOC 19/01/59.

WA521   F.1 specification. FF 11/51. AWC 20/11/51. Delivered RDU, Culdrose 21/11/51. 702 Sqn, Culdrose (no code) by 07/52. 736 Sqn, Culdrose (code 101/CW & 101/CU) 26/08/52. Moved with 736 Sqn to Lossiemouth (code 102/LM) 9/11/53-07/54. Airwork Ltd, Gatwick, Cat.4 for reconditioning by 01/55. Airwork, Dunsfold to AHU, Abbotsinch 16/05/55, Airwork, St.Davids (code 021/BY) 26/09/55. AHU, Abbotsinch 27/02/57. SFS 25/03/58.

WA522   F.1 specification. AWC 22/11/51. Delivered to RDU, Culdrose 26/11/51. 803 Sqn (code 119/J) 05/52. Noise heard prior to catapult launch, power loss, ditched ahead of HMS Eagle, Cat.ZZ, pilot (Lt A.W. Chandler USN) unhurt 10/07/52 [salt water in fuel was believed to be cause of accident]. SOC 28/07/52.

WA523   F.1 specification. AWC 30/11/51. Delivered RDU, Culdrose 4/12/51. 702 Sqn, Culdrose (code 194/CW) 28/03/52. 736 Sqn, Culdrose (code 104/CW & 104/CU) 26/08/52. Airwork Ltd, Gatwick, Cat.4 for reconditioning 5/09/53. Wheels up landing, Gatwick, Cat.H 11/03/55. Dunsfold to AHU, Abbotsinch for long term storage 23/08/55. SFS 25/03/58.

WA524   F.1 specification. AWC 30/11/51. Delivered to RDU, Culdrose 4/12/51. 890 Sqn 3/06/52. 736 Sqn, Culdrose 28/07/52. Hit by No.2 a/c while lining up for take-off 24/09/52. Ditched in sea off Porthleven, Cornwall after engine fire, Cat.ZZ, pilot (Lt(E) A.F, Brown) unhurt 25/03/53. SOC 30/03/53.

WA525 FB.1 specification. AWC 30/11/51. Delivered to RDU, Culdrose 10/12/51. To 800 Sqn (code 101/J) by 04/52. VA South Marston by 08/52. C Sqn, A&AEE for bombing trials 5/05/53. Airwork Ltd, Gatwick, Cat.4 for reconditioning 6/07/54. AHU, Abbotsinch 9/06/55. Airwork, St.Davids (code 017/BY) 17/01/56. AHU Abbotsinch 6/02/57. SFS 25/03/58.

WA526 FB.1 specification. FF 25/08/51. AWC 30/11/51. Delivered RDU, Culdrose 5/12/51. 703 Sqn, Ford 25/01/52. SOC, Ford 21/02/52. On dump at Ford by 09/52.

WA527 FB.1 specification. AWC 28/12/51. Delivered RDU, Culdrose 9/01/52. 703 Sqn, Ford (code 056/FD & 054/FD) 12/02/52. 800 Sqn, Ford (code 103/J) 19/04/52. AHU, Abbotsinch 12/12/52. 736 Sqn, Culdrose (code 119/CU) 21/10/53. 736 Sqn, Lossiemouth (code 119/LM) 4/11/53. Airwork Ltd, Gatwick, Cat.4 for reconditioning 27/08/54. AHU, Abbotsinch 1/11/55. SFS 25/03/58.

WA528 FB.1 specification. AWC 28/12/51, Delivered RDU, Culdrose 17/01/52. 800 Sqn, Ford (code 104/J) 31/03/52. AHU, Ford after tail oleo collapsed, no further flying 11/07/52. AHU, Abbotsinch by road 3/6/53, AHU, Ford by road 20/07/53. AHU, Abbotsinch by road for long term storage 24/08/53. Bramcote by road, GI Class II Instructional Airframe No.A2365 3/12/54. SOC 26/09/56. Scrap area 10/57.

WA529 FB.1 specification. FF 7/01/52. Vickers-Armstrong on permanent CS(A) charge 29/05/52. Sent by road and sea for winterisation trials at CE&PE (WEE) Namao, Canada 9/07/52. Rtnd to UK in HMCS Magnificent 05/53. Airwork, Gatwick to RAE Farnborough for arrester side load trials 24/07/54. GI Class II at AHU, Abbotsinch 27/10/54. RAE Bedford 7/03/56. Rtnd to RN charge 2/12/58. Ford for fire practice 21/06/58, later scrapped on site.

WA530 FB.1 specification. AWC 31/12/51. Delivered to RDU, Culdrose 16/01/52. 800 Sqn, Ford (code 105/J) 16/04/52. Forced landing at RAF Valley 7/09/52. Repaired on site by MARU, Stretton, work completed by 23/10/52. AHU, Abbotsinch by road 5/11/52. 736 Sqn, Culdrose (code 150/CU) 2/11/53. 736 Sqn, Lossiemouth (code 150/LM) 6/11/53. AHU, Abbotsinch for long term storage 14/07/54. Bramcote as GI Class II Instructional Airframe No.A2363 26/11/54. SOC 17/09/57 in scrapping compound by 10/57.

WA531 FB.1 specification. FF 1/01/52. AWC 8/01/52. Delivered RDU, Culdrose 9/01/52. 800 Sqn (code 102/J) 12/02/52. AHU, Abbotsinch 15/01/53. 738 Sqn, Lossiemouth (code 152/LM) 24/10/53-05/54. Airwork Ltd, Gatwick, Cat.4 for reconditioning 08/54. AHU, Abbotsinch 17/08/55. Airwork, St.Davids (code 020/BY) 17/10/55. AHU, Abbotsinch for long term storage 23/02/57. SFS 25/03/58.

WA532 FB.1 specification. AWC 29/01/52. Delivered RDU, Culham 4/02/52. 800 Sqn, Ford (code 106/J) 27/03/52. 890 Sqn, Ford (Code :106/J) 11/09/52. 800 Sqn, Ford (code 106/J) 11/10/52. AHU, Ford 22/01/53. Lee-on-Solent 13/03/53. Gosport 26/03/53. HMS Theseus for deck handling trials 30/03/53. Gosport 13/04/533. AHU, Abbotsinch by road 26/05/53. To HMS Implacable by Northern NATSU 14/07/53. To Gosport by road 21/07/53. AHU, Abbotsinch by road for HMS Perseus 4/08/53. To Bramcote by road, GI Class II Instructional Airframe No.A2366 3/12/54. SOC as scrap 17/09/57. Still in use 22/02/59. Last reported being used for fire practice at Brawdy in 1960.

WA533 FB.1 specification. AWC 29/01/52. Delivered RDU, Culdrose 5/02/52. 800 Sqn, Ford (code 107/J) 19/04/52. 890 Sqn, Ford (code 107/J) 15/09/52. 800 Sqn, Ford (code 107/J) 22/10/52. AHU, Abbotsinch for inspection 12/12/52. 736 Sqn, Culdrose (code 115/CU) 22/09/53. 736 Sqn, Lossiemouth (code 115/LM) 4/11/53. Both main flaps damaged, HMS Illustrious 17/05/54. AHU, Abbotsinch for long term storage 13/07/54. To Bramcote by road, GI Class II Instructional Airframe No.A2361 21/01/55. Seen in scrapping compound by 08/58. SOC 25/09/58.

WA534 FB.1 specification. AWC 31/01/52, Delivered RDU, Culdrose 5/02/52. 800 Sqn, Ford (code 108/J) 27/03/52. Forced landing after flame-out at 8,500ft 12/10/52. Engine failure, belly landing 5/11/52. AHU, Abbotsinch by road for inspection and repair 12/12/52. AHU, Culdrose 17/04/53. 736 Sqn, Culdrose (code 112/CW & 112/CU) 11/06/53: 736 Sqn, Lossiemouth (code 112/LM) 9/11/53, Airwork Ltd, Gatwick for overhaul 25/08/54. AHU, Abbotsinch for long term storage 8/12/55. SFS 30/05/58.

WA535 FB.1 specification. AWC 21/03/52. VA South Marston to C Sqn, A&AEE on loan for clearance of fully sealed elevator 31/03/52. VA Chilbolton 9/05/52. VA South Marston 4/06/52. AHU, Abbotsinch 9/10/52. 767 Sqn, Stretton 02/53. Made fast run over Burtonwood, canopy detached, made steep climb, then dived into a field at Winwick, nr Warrington, Cat.ZZ, pilot (Cdr Plt R E Collingwood) killed 5/02/53. Crash site excavated 13/09/2003, for exhibition at RAF Millom Museum.

WA536 Cancelled 2/01/51

WA537 Cancelled 2/01/51

Specification Number: E.1/45
Contract Number: 6/Acft/2822/CB.7(b) dtd 21/11/50
Type 511 FB.2
Quantity: 24

WK319 FF 25/04/52. AWC 20/05/52. To CS(A) VA Chilbolton 20/05/52. VA South Marston 5/08/52. Delivered C Sqn, A&AEE for R.P. and pylon bomb trials 7/10/52. VA South Marston 6/07/53. RAE Farnborough for catapult and arresting with bombs and R.P. trials 22/12/53. Airwork, Gatwick, Cat.4 for reconditioning 18/04/55. AHU, Abbotsinch for long term storage 15/03/56. SFS 25/03/58.

WK320 AWC 20/05/52. Delivered to RDU, Culdrose for long term storage 21/05/52. 890 Sqn, Ford (code 117/J, 143/J & 151/J) 18/07/52. Wheels-up landing at Milltown 22/10/52. HMS Eagle after repair 27/11/52. Precautionary landing on HMS Eagle 2/12/52. 803 Sqn, Ford (code 151/J) 8/12/52. Heavy landing on HMS Eagle 14/07/53. AHS, Ford 28/07/53. 736 Sqn, Lossiemouth 23/02/54. Unable to lock down starboard u/c, belly landed on runway, Cat.HX 23/02/54. AHU, Abbotsinch for repair 6/11/54. 1832 RNVR Sqn, Benson 23/05/56. 1833 RNVR Sqn, Honiley (code 833) 7/07/56. Arbroath 17/09/56. GI Class II Instructional Airframe No.A2418 16/10/56. SOC 19/01/59.

WK321 AWC 22/05/52. Delivered to RDU, Culdrose 3/06/52. 800 Sqn, Ford (code 111/J & 141/J) 23/07/52. 800 Sqn, Ford (code 111/J & 110/J) 8/12/52. Airwork, Gatwick, Cat.4 for reconditioning 4/06/54. Dunsfold to AHU, Abbotsinch 12/10/55. 1831 RNVR Sqn, Stretton (code 813/ST) 22/02/56. AHU, Lossiemouth for long term storage 30/01/57. SFS 5/03/58.

WK322 AWC 29/05/52. Delivered to RDU, Culdrose 3/06/52. 890 Sqn, Ford (code 112/J, 106/J & 142/J) 18/07/52. 800 Sqn, Ford (code 102/J) 6/12/52. Hit barrier on landing 4/03/53. Transferred to 803 Sqn for shipment from HMS Eagle to VA South Marston for repair 18/05/53. Transferred to Airwork, Gatwick by road for repair 14/04/53: AHU, Abbotsinch for long term storage 7/05/54. 718 Sqn, Stretton (code 1733/ST) 18/08/55. 1831 RNVR Sqn, Stretton (code 173/ST) 20/08/55. Port u/c failed to lower, belly landing on grass at Stretton, Cat.H 17/12/55. Airwork, Gatwick for repair 10/03/56. AHU, Abbotsinch 8/05/57. SFS 25/03/58.

WK323 AWC 29/05/52. Delivered to RDU, Culdrose 3/06/52. 890 Sqn, Ford (code 103/J) 28/10/52. 800 Sqn (code 105/J) 6/12/52. Engine fire during inverted flight, belly landing 11/12/52. ARS, Ford for repair. To Lossiemouth via HMS Eagle, still with 800 Sqn (code 102/J) 06/53. 800 Sqn, Ford 6/11/53. AHU, Abbotsinch for overhaul 28/06/54. 1832 RNVR Sqn, Benson (code 106 & 825) 11/10/55. AHU, Abbotsinch 15/10/56. SOC 25/03/58.

WK324 AWC 22/05/52. Delivered to RDU, Culdrose 22/05/52. 890 Sqn, Ford (code 114/J) 17/07/52. 800 Sqn (code 104/J) 09/52. Explosion and engine fire during flight from HMS Eagle, pilot (Lt C.R. Bushe) ejected, a/c ditched, Cat.ZZ 15/07/53. SOC 20/07/53.

WK325 AWC 22/05/52. Delivered to RDU, Culdrose 27/05/52. 890 Sqn, Ford (code 115/J) 22/07/52. Loaned to 803 Sqn on HMS Eaglc 9/09/52. 890 Sqn, Ford 11/10/52. Engine failure but landed safely at Lossiemouth 27/11/52. 800 Sqn, Ford (code 105/J) 3/12/52. Hit by Sea Hornet VW955 on HMS Eagle 10/01/54. Bounced over wires, then caught wire when entering barrier on HMS Eagle, Cat.L 28/04/54. Airwork, Gatwick, Cat.4 for reconditioning 4/06/54. AHU, Abbotsinch 7/07/55. 1832 RNVR Sqn, Benson (code 830) 30/04/56. 1831 RNVR Sqn, Stretton (code 830) 10/07/56. Port tailplane and main spar damaged after jettisoning R.P.s in sea, Cat.Y 9/08/56. Loaned to Airwork, St.Davids 11/09/56. AHU, Abbotsinch 21/09/56. SFS 25/03/58.

WK326 AWC 30/05/52. Delivered to RDU, Culdrose 4/06/52. 890 Sqn, Ford (code 116/J & 146/J) 21/07/52. 803 Sqn, Ford (code 175/J & 152/J) 11/06/54. Starboard u/c failed to lower and port leg raised before belly landing on ventral tank, Hal Far, Cat.SS 23/07/54. Airwork, Gatwick, Cat.3 for reconditioning 16/08/54. AHU, Abbotsinch 6/09/55. 1831 RNVR Sqn, Stretton (code 175/ST & 815/ST) 26/11/55. AHU, Lossiemouth 30/01/57. SFS 5/03/58.

WK327 AWC 30/05/52. Delivered to RDU, Culdrose 4/06/52. 890 Sqn, Ford (code 104 & 174) 26/08/52. 803 Sqn, Ford (code 146/J) 8/12/52. Engine fire after launch from HMS Eagle, successful emergency landing (rear cap on ventral tank found to be missing) 3/02/53. 800 Sqn, HMS Eagle (code 112/J & 108/J) 9/02/54. Airwork, Gatwick, Cat.4 for reconditioning 4/06/54. AHU, Abbotsinch 7/03/56. 1832 RNVR Sqn, Benson (code 850) 28/05/56. 1831 RNVR Sqn, Stretton (code 850) 8/07/56. AHU, Abbotsinch 29/10/56. SFS 25/03/57.

WK328 AWC 30/05/52. Delivered to RDU, Culdrose 4/06/52. 890 Sqn, Ford (code 118/J) 16/07/52. 803 Sqn, Ford (code 148/J) 8/12/52. Stn Flt, Ford 16/07/52. 736 Sqn, Lossiemouth 18/01/54. Airwork, Gatwick, Cat.4 for reconditioning 1/07/54. AHU, Abbotsinch 9/06/55. 1832 RNVR Sqn, Benson (code 828) 14/02/56. Aerodrome Dummy Deck Landing, flicked in on approach after being waved off, pilot (Lt J.S. Wyatt) killed, Ford, Cat.ZZ 25/06/56. SOC 19/07/56.

WK329 AWC 30/05/52. Delivered to RDU, Culdrose 9/06/52. 890 Sqn, Ford (code 113/J) 15/07/52. Loaned to 803 Sqn, stalled on final approach during Aerodrome Dummy Deck Landings, Ford 25/08/52. SOC 1/09/52. Remains seen at Arbroath in 03/55 (possibly for instructional use?)

WK330 AWC 8/08/52. Delivered to RDU, Culdrose 19/08/52. 890 Sqn, Ford 4/09/52. 803 Sqn, Ford (code 145/J). Engine fire on take-off, stalled into ground, cartwheeled and hit wall, pilot (Sqn Ldr A.W Griffin) killed, Hal Far, Cat. ZZ 25/05/54.

WK331 AWC 8/087/52. Delivered to RDU, Culdrose 20/08/52. 800 Sqn, Ford (code 107/J) 10/09/52. Heavy landing on HMS *Eagle* 24/09/53. AHU, Ford 6/10/53. 800 Sqn, Ford (code 107/J) 12/05/54. AHU, Abbotsinch 6/06/54. 1832 RNVR Sqn, Benson (code 103 & 822) 9/10/55. AHU, Abbotsinch 23/10/56. SFS 25/03/58.

WK332 AWC 24/09/52. Delivered to RDU, Culdrose 25/09/52. To 800 Sqn, Ford (code 106/J) by 11/52. Starboard u/c jammed up on landing, HMS *Eagle* 1/04/54. Floated over wires into barrier, HMS *Eagle*, Cat.H 12/05/54. Airwork, Gatwick by 07/54. SOC 15/01/55. Fuselage was sent to A&AEE for canopy jettison trials in blower tunnel.

WK333 AWC 8/08/52. Delivered to RDU, Culdrose 19/08/52. 787 Sqn, West Raynham 4/09/52. VA South Marston 1/12/52. AHU, Abbotsinch 21/07/53. 803 Sqn, HMS *Eagle* (code 143/J) 23/03/54. AHU, Hal Far by 08/54. AHU, Abbotsinch 17/08/54. Airwork, St.Davids (code 023/BY) 12/09/56. AHU, Abbotsinch 24/01/57. SFS 25/03/58.

WK334 AWC 8/08/52. Delivered to RDU, Culdrose 20/08/52. 803 Sqn, Ford (code 142). Collided with WK339 at 4,000ft north of Manhood Range (Selsey Bill) during air to ground firing, crashed Sidlesham, Chichester, pilot (Lt F D B Bailey) killed, Cat.ZZ 19/05/53. SOC 22/05/53.

WK335 AWC 26/09/52. Delivered to RDU, Culdrose 29/09/52. 800 Sqn, Ford 31/10/52. Fuel cap cover missing, fuel leaking, a/c caught fire and exploded at 1,5000ft during fly-past over Yugoslav training ship Galeb, dived into sea off Gibraltar, pilot (Lt P.G. Ree) did not eject, killed and body never recovered, Cat.ZZ 11/03/53. SOC 27/03/53.

WK336 AWC 26/09/52. Delivered to RDU, Culdrose 29/09/52. AHU, Ford 27/10/52. 800 Sqn, Ford (code 109/J) 26/01/53. Heavy landing 19/06/53. ARS, Ford 27/07/53. Airwork, Gatwick by road for repair 21/08/52. AHU, Abbotsinch 7/05/54. 718 Sqn, Honiley (code 164) 22/09/55. 1833 RNVR Sqn, Honiley (code 164 & 839) 24/10/55. AHU, Lossiemouth 12/02/57. SOC as scrap 10/57.

WK337 AWC 26/09/52. Delivered to AHU, Abbotsinch 7/10/52. 800 Sqn, Ford (code 110/J) 11/12/52. Engine fire in inverted flight, wheels-up landing, HMS *Eagle* 5/05/53. Airwork, Gatwick for reconditioning 23/07/53. AHU, Abbotsinch via Ford 2/02/55. Airwork, FRU Hurn (code 036 [applied 01/56]) 6/12/55. Jet pipe fire on start-up, Cat. LY 10/09/56. Flown to AHU, Abbotsinch 20/02/57. Put up for sale as scrap 25/03/58. SOC 25/04/58.

WK338 AWC 3/10/52. Delivered to AHU, Abbotsinch 9/10/52. 800 Sqn, Ford (code 108/J) 12/12/52-01/54. Airwork, Gatwick for reconditioning by 5/05/55. AHU, Abbotsinch 3/06/55. 1832 RNVR Sqn, Benson (code 104 & 823) 9/10/55. Stalled, heavy landing on runway, spar fractured, Cat. HY 9/05/56. Airwork, Gatwick by road 31/05/56. SOC 24/09/58 and reduced to components.

WK339 AWC 26/09/52. Delivered to AHU, Abbotsinch 8/10/52. 800 Sqn, Ford 21/04/53. Collided with WK334 at 4,000ft north of Manhood Range (Selsy Bill) during air to ground firing, pilot (Lt Cdr J.M. Glaser) killed, Cat.ZZ 19/05/53. SOC 22/05/53.

WK340 AWC 14/11/52. Delivered to AHU, Abbotsinch 24/11/52. 803 Sqn (code 141/J) by 06/53. Engine caught fire during take-off from HMS *Eagle*, pilot (Lt D.I. Berry) ejected at 300ft and was killed, Cat.ZZ 2/04/54. SOC 29/04/54. Tail section salvaged and recovered to Malta.

WK341 AWC 14/11/52. Delivered to AHU, Abbotsinch 24/11/52. 800 Sqn, Ford (code 112/J) 19/01/53. 800 Sqn, HMS *Eagle* (code 112/J) 9/02/53. ARS, Ford 27/07/53. Airwork, Gatwick by road for repair 21/08/53. AHU, Abbotsinch 26/08/54. 718 Sqn, Stretton (code 171/ST) 19/06/55. 1831 RNVR Sqn, Stretton (code 171/ST & 711/ST) 4/07/55. AHU, Lossiemouth 28/01/57. SOC 5/03/58.

WK342 AWC 21/11/52. Delivered to AHU, Abbotsinch 24/11/52. 803 Sqn, Ford (code 144/J) 21/11/52. Airwork, Gatwick, Cat.4 for repair 30/09/53. AHU, Abbotsinch 16/02/55. Airwork, FRU Hurn (code 034 [applied 01/56]) 8/11/55. Flown to AHU, Abbotsinch for long term storage 12/02/57. Put up for sale as scrap 25/03/58. SOC 25/04/58.

Specification Number: E.1/45
Contract Number: 6/Acft/6343/CB.7(b) dtd 16/02/51
Type 511 FB.2
Quantity: 30

WP275 AWC 31/12/52. Delivered to AHU, Abbotsinch 3/02/53. AHU, Ford 25/03/53. 800 Sqn, Ford (code 102/J) 23/04/53. 803 Sqn, Ford (code 112/FD) 27/07/53. Heavy landing, starboard u/c collapsed 27/11/53. Airwork, Gatwick by road, Cat.4 for repair 7/12/53. AHU, Abbotsinch 2/06/55. SAD, Benson (code 827) 8/02/56. Wing tip folded after take-off, pilot (Sqn Ldr J.F. Yeates) ejected safely, a/c crashed in sea, Cat.ZZ 6/07/56. WOC 19/07/56.

WP276 AWC 21/11/52. Delivered to AHU, Abbotsinch 24/11/52. 803 Sqn, Ford (code 145/J) 22/12/52. Hook pulled out, into barrier, HMS *Eagle* 26/01/53. ARS, Ford for repair 29/01/53. VA South Marston 25/03/53. AHU, Abbotsinch 11/05/53. 718 Sqn, Stretton (code 174) 19/08/55. 1831 RNVR Sqn, Stretton (code 174 & 814/ST) 20/080/55. Power loss, precautionary landing, Cat.SS 27/08/55. Tailwheel collapsed at end of landing run, Cat. LQ 5/10/56. AHU, Lossiemouth for long term storage 14/02/57. WOC and SFS 10/57.

WP277 AWC 2/12/52. Delivered to AHU, Abbotsinch 21/01/53. AHU, Hal Far 20/02/53. 803 Sqn, HMS *Eagle* (code 154/J & 145/J) 15/03/53. Brakes failed during taxying, ran into WP292, HMS *Eagle*, Cat. LC 5/03/54. AHU, Hal Far 11/11/54. AHU, Abbotsinch via Lee-on-Solent for inspection and storage 29/03/55. A&AEE by road for trials and intensive inspection 07/56. AHU, Abbotsinch for long term storage 15/12/56. Cocooned 21/01/57. SFS 25/03/58.

WP278 AWC 31/12/52. Delivered to AHU, Abbotsinch 21/01/53. Rolls-Royce, Hucknall for relighting trials 28/03/53. AHU, Abbotsinch for long term storage 19/08/54. SFS 25/03/58.

WP279 AWC 31/10/52. Delivered to AHU, Abbotsinch 10/11/52. 803 Sqn, Ford (code 147/J) 14/01/53. Flame-out, emergency wheels-up landing, Krendi, Cat.SS 3/06/54. AHU, Hal Far 15/06/54. Rtnd to UK. Airwork, Gatwick, Cat.4 for repair 17/09/54. AHU, Abbotsinch 2/11/55. SAD, Benson (code 829) 23/04/56. 1833 RNVR Sqn, Honiley (code 829) 7/07/56. AHU, Abbotsinch 17/09/56. SFS 25/03/58.

WP280 FF 28/10/52. AWC 31/10/52. Delivered to AHU, Abbotsinch 10/11/52. 803 Sqn, Ford (code 149/J) 16/01/53. ARS, Ford for inspection 28/07/53. 800 Sqn, HMS *Eagle* (code 104/J) 16/11/53. AHU, Hal Far 5/05/54. AHU, Abbotsinch for inspection and repair 15/05/54. SAD, Benson (code 828) 26/06/56. 1833 RNVR Sqn, Honiley (code 828) 7/07/56. AHU, Abbotsinch 17/09/56. SFS 25/03/58.

WP281 AWC 12/12/52. Delivered to AHU, Abbotsinch 2/02/53. 718 Sqn, Stretton (code 172/ST) 11/07/53. 1831 RNVR Sqn, Stretton (code 172/ST) 18/07/55. Lost control after evasive action to avoid Sea Prince after take-off, dived into ground east of the airfield boundary, pilot (Lt Cdr C.J. Lavender) killed, Cat.ZZ 10/11/55. SOC 16/11/55.

WP282 AWC 2/12/52. Delivered to AHU, Abbotsinch 21/01/53. 787 Sqn, West Raynham 25/02/53. AHU, Abbotsinch for long term storage 20/08/54. SFS 25/03/58.

WP283 AWC 19/12/52. Delivered to AHU, Abbotsinch 2/02/53. 736 Sqn, Culdrose (code 117/CU) 29/10/53. With 737 Sqn when it moved to Lossiemouth 9/11/53. AHU, Abbotsinch 13/07/54. 718 Sqn, Honiley (code 166/ST) 13/10/55. 1833 RNVR Sqn (code 166/ST & 841/ST) 24/10/55. Stalled during Aerodrome Dummy Deck Landings, killed civilian on nearby road and injured another, pilot (Sqn Ldr R.M. Neill) unhurt. Cat.ZZ 15/07/56. SOC 25/07/56.

WP284 AWC 2/12/52. Delivered to AHU, Abbotsinch 2/02/53. NARIU Gosport 19/06/53. AHU, Abbotsinch 6/10/53. 800 Sqn, Ford (code 103/J) 11/12/53. AHU, Abbotsinch 16/06/54. 1832 RNVR Sqn, Benson (code 823) 1/06/56. AHU, Abbotsinch 2/07/56. SFS 25/03/58.

WP285 AWC 29/12/52. Delivered to AHU, Abbotsinch 4/02/53. RN Handling Sqn, Manby 23/04/53. 736 Sqn, Lossiemouth 10/05/54. AHU, Abbotsinch 13/07/54. Airwork, FRU Hurn (code 038) 18/01/56. Flown to AHU, Abbotsinch for long term storage 6/02/57. For sale as scrap 25/03/58. SOC 25/04/58.

WP286 AWC 31/12/52. Delivered to AHU, Abbotsinch 3/02/53. 800 Sqn, Ford (code 101/J) 28/03/53. AHU, Abbotsinch 10/06/54. 718 Sqn, Honiley (code 165) 22/09/55. 1833 RNVR Sqn, Honiley (code 165 & 840) 24/10/55. AHU, Lossiemouth for long term storage 12/02/57. Arbroath as GI Class II Instructional Airframe No.A2446 2/10/57. WOC 19/01/59.

WP287 AWC 31/12/52. Delivered to AHU, Abbotsinch 3/02/53. 800 Sqn, Ford 8/06/53. 703 Sqn, Ford 15/06/53. 800 Sqn, HMS *Eagle* (code 110/J & 832/ST) 6/07/53. AHU, Abbotsinch 7/10/54. SAD, Benson (code 824 & 832/ST) 16/05/56. NAD, Stretton 10/07/56. AHU, Abbotsinch 21/09/56. SFS 25/03/58.

WP288 AWC 31/01/53. Delivered AHU, Abbotsinch 6/02/53. 803

Sqn, Ford by 07/53. Mid-air collision with WZ296 during unauthorised low level formation aerobatics, a/c crashed 9 miles SW of Filfa Island, pilot (Sqn Ldr G R Finch) killed, Hal Far, Cat.ZZ 20/08/54.

WP289 AWC 31/01/53. Delivered to AHU, Abbotsinch 4/02/53. AHU, Ford 20/02/53. 800 Sqn, Ford 27/02/53. Airwork, Gatwick for repair by road 15/07/53. AHU, Abbotsinch 1/02/55. 1832 Sqn RNVR, Benson (code 107 & 826) 17/10/55. AHU, Abbotsinch 31/10/56. SFS 25/03/58.

WP290 AWC 31/01/53. Delivered to AHU, Abbotsinch 4/02/53. 787 Sqn, West Raynham 26/02/53. Fuel transfer failure, wheels-up landing on beach south of Mablethorpe, Lincs, Cat.LX 16/03/54. Airwork, Gatwick by road for repair 29/04/54. VA South Marston to AHU, Abbotsinch 1/02/55. 1831 RNVR Sqn, Stretton (code 812/ST) 8/02/56. AHU, Lossiemouth for long term storage 28/01/57. WOC 10/57.

WP291 AWC 31/01/53. Delivered to AHU, Abbotsinch 4/02/53. 787 Sqn, West Raynham 26/02/53. Rolls-Royce, Hucknall for investigation into engine roughness 17/11/54. Airwork, Gatwick for reconditioning 20/04/55. AHU, Abbotsinch 16/04/56. SFS 25/03/58.

WP292 AWC 31/01/53. Delivered to AHU, Abbotsinch 6/02/53. 787 Sqn, West Raynham 26/02/53. 803 Sqn, Ford (code 143/J) 4/06/53. By road to HMS Implacable for trials 25/03/54. AHU, Yeovilton by road for repair by MARU 6/054/54. AHU, Abbotsinch for long term storage 16/07/54. SFS 25/03/58.

WP293 AWC 18/02/53. Delivered to AHU, Abbotsinch 23/02/53. 703 Sqn, Ford 1/07/53. Familiarisation flight, a/c seen to be climbing and diving, sometimes inverted, flew into ground North Stoke Farm, Houghton, Nr Arundel, Sussex, pilot (Lt W T R Smith) killed, Cat.ZZ 21/07/53. Hydraulic fluid was discovered in canopy during investigation.

WP294 AWC 25/02/53. Delivered to AHU, Abbotsinch 26/02/53. 803 Sqn, Ford (code 153/J & 142/J) 4/05/53. MARU, Ford for repairs 2/11/53. AHU, Abbotsinch for long term storage 6/12/54. SFS 25/03/58.

WP295 AWC 26/02/53. Delivered to AHU, Abbotsinch 11/03/53. 803 Sqn, Ford (code 149/J) 8/06/53. HMS Eagle, lost control during catapult launch, turned over and went into sea off Gibraltar, pilot (Lt J F Nash) killed, Cat.ZZ 22/02/54.

WP296 AWC 26/02/53. Delivered to AHU, Abbotsinch 11/03/53. 767 Sqn, Stretton (code 171/JA & 171/ST) 10/07/53. 736 Sqn, Lossiemouth (code 157/LM) 18/03/54. AHU, Abbotsinch for long term storage 22/06/54. SFS 25/03/58.

WP297 AWC 26/02/53. Delivered to AHU, Abbotsinch 11/03/53. 800 Sqn, Ford 4/06/53. 703 Sqn, Ford 15/06/53. 800 Sqn, Ford (code 104/J) 20/07/53. Heavy landing 2/11/53. ARS, Ford 12/11/53. Airwork, Gatwick by road for repair 18/11/533. AHU, Abbotsinch 2/02/55. Airwork, FRU Hurn (code 035 [NB. No confirmation that this aircraft actually ever had this number applied to it]) 29/11/55. Port u/c would not lower. Belly landing on airfield, Cat. LY 30/11/56, RNAY Fleetlands 15/03/57. RSP 4/03/58. SOC 26/06/58.

WP298 AWC 28/02/53. Delivered to AHU, Abbotsinch 12/03/53. 800 Sqn, Ford (code 102/J) 20/05/53. Heavy landing 22/09/53. AHU, Lossiemouth for repair 29/09/53. 800 Sqn, Ford (code 102/J) 15/10/53. AHU, Abbotsinch 19/07/54. SFS 25/03/58.

WP299 AWC 3/03/53. Delivered to AHU, Abbotsinch 12/03/53. AHU, Culdrose 7/05/53. 736 Sqn, Culdrose (code 109/CW) 27/05/53. With 736 Sqn to Lossiemouth (code 109/LM) 9/11/53. AHU, Abbotsinch for long term storage 13/07/54. SFS 25/03/58.

WP300 AWC 9/03/53. Delivered to AHU, Abbotsinch 12/03/53. 803 Sqn, Ford (code 151/J) 27/05/53. Power failure, crashed short of runway, Hal Far, Cat.HY 2/04/54. SOC 26/04/54.

WP301 AWC 20/03/53. Delivered to AHU, Abbotsinch 25/03/53. 767 Sqn, Stretton 15/07/53. 736 sqn, Lossiemouth (code 154/LM) 22/03/54. AHU, Abbotsinch 13/07/54. SAD, Benson (code 108 & 827) 20/11/55. AHU, Abbotsinch 17/03/56. SFS 25/03/58.

WP302 AWC 20/03/53. Delivered to AHU, Abbotsinch 24/03/533. 803 Sqn, Ford (code 153/J) 4/06/53. AHU, Ford 29/07/53. NARIU. Gosport (later moving to Lee-on-Solent) 13/01/54. AHU, Abbotsinch 9/07/54. 1833 RNVR Sqn, Honiley (code 836) 14/08/56. AHU, Abbotsinch for long term storage 24/01/57. SFS 25/03/58.

WP303 AWC 20/03/53. Delivered to AHU, Abbotsinch 25/03/53. 803 Sqn, Ford (code 150/J) 25/09/53. AHU, Abbotsinch for long term storage 17/09/56. SFS 25/03/58.

WP304 AWC 27/03/53. Delivered to AHU, Abbotsinch 16/04/53. 803 Sqn, Ford (code 109/J) 23/06/53. Missed wires, hit barrier, Cat. H 17/03/54. Stripped for spares and ditched over the side, HMS Eagle 29/03/54. SOC 7/04/54.

Specification Number: E.1/45
Contract Number: 6/Acft/7319/CB.7(b) dtd 7/09/51
Type 511 FB.2
Quantity: 30

WZ273 AWC 31/03/53. Delivered to AHU, Abbotsinch 16/04/53.

767 Sqn, Stretton (code 170/ST) 15/07/53. 736 Sqn, Lossiemouth (code 155/LM) 22/03/54. AHU, Abbotsinch for long term storage 15/07/54. 1832 RNVR Sqn, Benson (code 823) 27/06/56. AHU, Abbotsinch for long term storage 31/10/56. SFS 25/03/58.

WZ274 AWC 31/03/53. Delivered to AHU, Abbotsinch 6/05/53. 737 Sqn, Culdrose (code 108/CU0 14/10/53. With 736 Sqn to Lossiemouth (code 106/LM) 4/11/533. AHU, Abbotsinch for long term storage 13/07/54. SAD, Benson (code 834) 23/05/56. 1833 RNVR Sqn, Honiley (code 834) 7/07/56. AHU, Abbotsinch for long term storage 24/09/56. SFS 25/03/58.

WZ275 AWC 31/03/53. Delivered to AHU, Abbotsinch 23/04/53. 767 Sqn, Stretton (code 172/ST) 13/07/53. 736 Sqn, Lossiemouth (code 153/LM) 19/03/54. AHU, Abbotsinch for long term storage 14/07/54. SFS 25/03/58.

WZ276 FF 28/04/53. AWC 30/04/53. Delivered to AHU, Abbotsinch 6/05/53. 800 Sqn, Ford (code 111/J) 24/08/53. Floated over wires and hit barrier, HMS Eagle, Cat/H 2/02/54. Airwork, Gatwick by road 6/02/54. To Gosport by road 10/03/55. SOC 23/03/55.

WZ277 AWC 30/04/53. Delivered to AHU, Abbotsinch 8/05/53. 703 Sqn, Ford (code 076/FD) 1/04/55. 700 Sqn, Ford (code 076/FD) 18/08/55. AHU, Abbotsinch 15/02/56. SOC 25/03/58.

WZ278 AWC 30/04/53. Delivered to AHU, Abbotsinch 8/03/53. 803 Sqn, Ford (code 148/J) 24/05/53. AHU, Hal Far 11/11/54. Ferried to UK 1/02/55. Lee-on-Solent 3/02/55. AHU, Abbotsinch for long term storage 17/02/55. SFS to Minworth Metal 30/05/58.

WZ279 AWC 30/04/53. Delivered to AHU, Abbotsinch 7/05/53. 800 Sqn, Ford (code 103/J) 25/08/53. Heavy landing 27/11/53. ARS, Yeovilton 14/12/53. AHU, Abbotsinch 7/03/55. Arbroath 13/12/55. To GI Class II Instructional Airframe No.A2400 9/01/56. SOC 29/01/59.

WZ280 AWC 30/04/53. Delivered to AHU, Abbotsinch 11/05/53. 736 Sqn, Culdrose (code 110/CU) 5/10/53. With 736 Sqn to Lossiemouth (code 110/LM) 9/11/53. AHU, Abbotsinch for long term storage 13/07/54. SFS 25/03/58.

WZ281 AWC 29/05/53. Delivered to AHU, Abbotsinch 3/06/53. 803 Sqn, Ford (code 147/J) 24/08/53. AHU, Abbotsinch long term storage 16/08/54. 1833 RNVR Sqn, Honiley (code 838) 16/11/56. AHU, Lossiemouth 14/02/57. SFS 5/08/58.

WZ282 AWC 29/05/53. Delivered to AHU, Abbotsinch 3/06/53. 800 Sqn, Ford (code 105/J) 25/06/53. AHU, Ford for repair 6/11/53. AHU, Lossiemouth 5/04/54. 736 Sqn, Lossiemouth (code 105/LM) 18/06/54. AHU, Abbotsinch 13/07/54. NARIU, Lee-on-Solent 9/03/56. AHU, Abbotsinch 17/10/56. SFS 25/03/58.

WZ283 AWC 29/05/53. Delivered to AHU, Abbotsinch 3/06/53. 718 Sqn, Stretton (code 170/ST) 18/06/55. 1831 RNVR Sqn, Stretton (code 170/ST & 810/ST) 4/07/55. To AHU, Lossiemouth by 03/57. SFS 5/03/58.

WZ284 AWC 30/05/53. Delivered to AHU, Abbotsinch 4/06/53. 800 Sqn (code 109/J), crashed 300yds short of runway, Hal Far, Cat.HZ 15/07/54. SOC 29/07/54.

WZ285 AWC 30/05/53. Delivered to AHU, Abbotsinch 4/06/53 - 7/53. 736 Sqn, Lossiemouth by 10/53. Lost height during Aerodrome Dummy Deck Landings, power applied late, struck brick wall short of runway and flew into ground, pilot (Lt R.J. Duncan) unhurt, Cat.ZZ 9/02/54. SOC 23/02/54.

WZ286 Taken on charge by CS(A) at VA South Marston 17/06/53. AWC 30/06/53. VA Chilbolton 18/09/53. VA South Marston for tests 27/05/54. AHU, Abbotsinch 21/06/54. Airwork, FRU Hurn (code 033) 1/11/55. AHU, Abbotsinch for long term storage 26/02/57. SFS 25/03/58.

WZ287 AWC 15/07/53. Delivered to AHU, Abbotsinch 21/07/53. 800 Sqn, Ford (code 102/J) 10/11/53. Hook pulled out when arrester gear jammed, HMS Eagle, Cat.L 6/04/54. Cleared for one flight only, u/c locked down for flight to Airwork, Gatwick, Cat.4 for repair 1/06/54. AHU, Abbotsinch 8/03/55. SAD, Benson (code 105 & 824) 5 /09/55. To AHU, Lossiemouth 09/56. SFS 25/03/58.

WZ288 AWC 31/07/53. Delivered to AHU, Abbotsinch 18/08/53. 800 Sqn, Ford (code 105/J) 12/11/53. AHU, Abbotsinch 5/06/54. SAD, Benson (code 102 & 821) 10/10/55. AHU, Abbotsinch 5/09/56. SFS 25/03/58.

WZ289 AWC 31/07/53. Delivered to AHU, Abbotsinch for long term storage 18/08/53. 1831 RNVR Sqn, Stretton (code 175/ST) 8/09/55. Low pull-out from strafing dive, stalled, wing buckled, Cat.L 30/10/55. To Arbroath by road 13/12/555. GI Class II Instructional Airframe No.A2401 14/01/56. WOC 19/01/57.

WZ290 AWC 31/07/53. Delivered to AHU, Abbotsinch 14/08/53. 803 Sqn, Ford (code 149/J) 8/12/53. AHU, Abbotsinch for long term storage 17/09/54. SFS 25/03/58.

WZ291 AWC 31/07/53. Delivered to AHU, Abbotsinch 14/08/53. 800 Sqn, Ford (code 112) 19/12/53. AHU, Abbotsinch 10/06/54. Airwork, FRU Hurn (code 037) 17/01/56. AHU, Abbotsinch for long term storage 13/02/57. SFS 25/03/58.

WZ292   AWC 31/08/53. Delivered to AHU, Abbotsinch 3/09/53. 800 Sqn, Ford (code 111 & 112/J) 9/12/53. AHU, Abbotsinch 10/06/54. 1833 RNVR Sqn, Bramcote 6/12/55. With 1833 RNVR Sqn to Honiley (code 163 & 828/ST) 10/12/55. Lift spoiler jammed on take-off, failed to retract, a/c failed to climb after take-off, touched down and u/c retracted, crossed road and ended up in road. Cat.HY 3/11/56. By road to Lee-on-Solent 12/11/56. RNAY Fleetlands and RSP 26/03/57.

WZ293   AWC 31/08/53. Delivered to AHU, Abbotsinch 3/09/53. 800 Sqn, HMS *Eagle* 11/05/54. 803 Sqn (code 157/J) 16/05/54. AHU, Abbotsinch 16/09/54. SAD, Benson (code 831) 4/05/56. NAD, Stretton (code 831/ST) 10/07/56. AHU, Abbotsinch 21/09/56. SFS 25/03/58.

WZ294   AWC 31/08/53. Delivered to AHU, Abbotsinch 8/09/53. HMS *Eagle* 24/03/54. 800 Sqn, North Front (RAF Gibraltar) 26/03/54. AHU, Abbotsinch 10/06/54. 718 Sqn, Stretton (code 176/ST) 7/09/55. 1831 RNVR Sqn (code 176/ST & 816/ST) 10/09/55. AHU, Lossiemouth 30/01/57. SFS 5/03/58.

WZ295   AWC 30/09/533. Delivered to AHU, Abbotsinch 7/10/53. 803 Sqn, Hal Far (code 140/J) 14/02/54. Rtnd to UK, arrived Lee-on-Solent 17/08/54. AHU, Abbotsinch by 01/55-7/56. 1833 RNVR Sqn, Honiley (code 160/BR & 841) by 10/56. AHU, Lossiemouth 12/02/57. SFS 5/03/58.

WZ296   AWC 30/09/533. Delivered to AHU, Abbotsinch 6/10/53. 803 Sqn, Hal Far 1/07/54. Mid-air collision with WP288 during unauthorised low-level formation aerobatics, flew into sea 9 miles SW of Filfa Island, pilot (Lt(E) J H Stock) killed, Cat.ZZ 20/08/54.

WZ297   AWC 30/09/53. Delivered to AHU, Abbotsinch 6/10/53. 787 Sqn, West Raynham 9/04/54. Airwork, St.Davids (code 022/BY) 12/09/56. AHU, Abbotsinch for long term storage 22/01/57. SFS 25/03/58.

WZ298   AWC 31/10/53. Delivered to AHU, Abbotsinch 3/11/53. AMCO, Lee-on-Solent 9/04/54. AHU, Hal Far 12/04/54. 800 Sqn, Hal Far (code 102/J) 23/04/54. AHU, Abbotsinch via Ford for long term storage 10/06/54. SFS 25/03/58.

WZ299   AWC 6/11/53. Delivered to AHU, Abbotsinch for long term storage 11/11/53. 718 Sqn, Stretton (code 160/ST) 6/05/55. 1833 RNVR Sqn, Honiley (code 160 & 835) 24/10/55. AHU, Lossiemouth for long term storage 14/02/57. To Arbroath by road 2/10/57. GI Class II Instructional Airframe No.A2445 7/10/57. WOC 19/01/59.

WZ300   AWC 9/11/53. Delivered to AHU, Abbotsinch 11/11/53. AHU, Stretton 2/06/55. 718 Sqn (code 161/ST) 4/06/55. 1833 RNVR Sqn, Honiley (code 161 & 836) 24/10/55. Overshot landing at Ford, Cat.HY 10/08/56. ARS, Ford 13/08/56. Airwork, Gatwick by road for repair 14/09/56. Declared Cat.5 and reduced to spares 24/09/58.

WZ301   AWC 31/12/53. Delivered to AHU, Abbotsinch 14/11/54. 718 Sqn, Stretton (code 162/ST) 13/05/55. 1833 RNVR Sqn, Honiley (code 162 & 837) 24/10/55. Party removed drogue gun from ejection seat of WP286 fired and pierced fuselage, of WZ301. Cat.LC 30/01/56. AHU, Lossiemouth 11/02/57. SFS 5/03/58.

WZ302   AWC 4/03/54. Delivered to AHU, Abbotsinch 8/03/54. 718 Sqn, Stretton (code 163/ST) 16/05/55. To 1833 RNVR Sqn, Honiley but stalled on approach, crashed Wroxall, 1 mile from airfield, pilot (Lt W Fletcher) killed, Cat.ZZ 15/10/55. Reduced to spares and SOC 1/11/55.

Specification Number: E.1/45
Contract Number: 6/Acft/2822/CB.7(b) dtd 27/03/51
Type 511 FB.1
Quantity: 1

WT851   Built to replace WA477. Delivered RDU, Culdrose for long term storage 16/05/52. 736 Sqn, Culdrose (code 111/CW & 111/CU) 26/09/52. Lossiemouth (code 111/LM) 9/11/53. Airwork Ltd, Gatwick (later moved to Lasham), Cat.4 for reconditioning 21/07/54. AHU, Abbotsinch 8/11/56. SOC and SFS 25/03/58.

Specification Number: N/A
Contract Number: N/A
Type 538 De-navalised FB.2s w/o wing fold
Quantity: 36, allocated serial numbers R4000 to R4035. Delivered between June 1951 and May 1953 to Drigh Road, Karachi.

F.1 WA473 is the last surviving example of the Attacker. Seen here at the FAA Museum at Yeovilton, where it was photographed by the author enjoying a rare and brief outdoor excursion during the airshow at the base in the late 1980s.
Authors collection

GLOSSARY AND BIBLIOGRAPHY

A&AEE	Aeroplane & Armament Experimental Establishment
AFB	Air Force Base
AHU	Aircraft Handling Unit
AIU	Accident Investigation Unit
AOC	Air Officer Commanding
ARS	Aircraft Repair Section
A.S.	Anti-Submarine
ASR	Air-Sea Rescue
ASV	Air to Surface Vessel (Radar)
ASWDU	Air-Sea Warfare Development Unit
AWC	Awaiting Collection
BS	British Standard
C(A)	Controller (Aircraft)
Capt	Captain
CFS	Central Flying School
CinC	Commander-in-Chief
CO	Commanding Officer
CRD	Civilian Repair Depot
CS(A)	Controller of Supplies (Air). Changed to C(A) 05/54
CSDE	Central Servicing Development Establishment
DF	Direction Finding
DFC	Distinguished Flying Cross
DFM	Distinguished Flying Medal
DSO	Distinguished Service Order
DTD	Directorate of Technical Development
FAA	Fleet Air Arm
FB	Fighter-bomber (role prefix)
FF	First Flight
Fg Off	Flying Officer (RAF)
Flt	Flight
Flt Sgt	Flight Sergeant (RAF)
Flt Lt	Flight Lieutenant (RAF)
FRU	Fleet Requirements Unit
FS	Federal Standard
ft	Foot
g	Acceleration of free fall due to gravity
GHQ	General Headquarters
Gp Capt	Group Captain (RAF)
H.E.	High Explosive
HMS	His/Her Majesty's Ship
HQ	Headquarters
IAM	Institute of Aviation Medicine
IFF	Identification Friend or Foe
in.	Inch
JASS	Joint Anti-Submarine School
kg	Kilogram
km	Kilometre
km/h	Kilometres Per Hour
knot	Unit of speed of 1nm per hour (approx 1.15mph or 1.85km/h)
lb	Pound
lt	Litre
Lt	Lieutenant
Lt Cdr	Lieutenant Commander (Royal Navy)
Maj	Major
MAP	Ministry of Aircraft Production
MARU	Mobile Aircraft Repair Unit
MEAF	Middle East Air Force
Mk	Mark
MofA	Ministry of Aviation
MOTU	Maritime Operational Training Unit
mph	Miles Per Hour
MTPS	Maintenance Test Pilots School
MU	Maintenance Unit (RAF)
NAD	Naval Air Department (RAE Farnborough)
NAFDU	Naval Air Fighting Development Unit
NAMDU	Naval Aircraft Maintenance Development Unit
NARIU	Naval Air Radio Installation Unit
No.	Number
NCO	Non-commissioned Officer
OC	Officer Commanding
OCU	Operational Conversion Unit
OTU	Operation Training Unit
Plt Off	Pilot Officer (RAF)
RAE	Royal Aircraft Establishment
RAF	Royal Air Force
RAE	Royal Aircraft Establishment
RDU	Receipt & Despatch Unit
RN	Royal Navy
RNAY	Royal Naval Aircraft Yard
RNAS	Royal Naval Air Station
RNVR	Royal Navy Volunteer Reserve
R.P.	Rocket Projectile
RRE	Radar Research Establishment
RSP	Reduced to spares & produce
RSU	Repair & Servicing Unit
SAD	Southern Air Division (RNVR)
SAR	Search & Rescue
SBAC	Society of British Aircraft Constructors
SFS	Sold for scrap
Sgt	Sergeant
SOC	Struck Off Charge
Sqn	Squadron
Sqn Ldr	Squadron Leader (RAF)
UK	United Kingdom
USAF	United States Air Force
USN	United States Navy
USS	United States Ship
VA	Vickers-Armstrong
VC	Victoria Cross
Wg Cdr	Wing Commander (RAF)
W/O	Warrant Officer
WOC	Written off charge
W/T	Wireless Telegraphy
WWI	World War I (1914-1918)
WWII	World War II (1939-1945)

British Secret Projects - Jet Fighters since 1950
Tony Buttler. Published by Midland Publishing ©2000
ISBN: 1-85780-095-8

Combat Aircraft Since 1945
S Wilson. Published by Airlife ©2001

Farnborough - 100 Years of British Aviation
Peter J Cooper. Midland Publishing Ltd ©2006
ISBN: 1-85780-239-X

Fleet Air Arm Fixed-Wing Aircraft since 1946
R Sturtivant. M Burrow and L Howard
Published by Air Britain ©2004
ISBN: 0-85130-283-1

Supermarine Aircraft since 1914
C F Andrews and E B Morgan. Published by Putnam ©1981
ISBN: 0-370-10018-2

Supermarine Attacker, Swift and Scimitar
P Birtles. Postwar Military Aircraft No7. Published by Ian Allan 1992
ISBN: 0-7110-2034-5

Spitfire - A Test Pilot's Story
Jeffrey Quill OBE, AFC, FRAeS. Published by Arrow Books 1983
and 1989
ISBN: 0-09-937020-4

The Cold War Years - Flight Testing at Boscombe Down 1945-1975
Tim Mason. Published by Hikoki Publications 2001
ISBN: 1-902109-11-2

The Squadrons of the Fleet Air Arm
Ray Sturtivant and Thoe Ballance. Air-Britain Publication 1994
ISBN: 0-85130-223-8

Official Publications
Air Ministry/MoD Air Publications
F.1 - Air Publication 4302A
FB.1 - Air Publication 4302A
FB.2 - Air Publication 4302A/B
Air Recognition Journal Vol.4, No.2 October 1949

Periodicals
Aeromodeller July 1947
Aeroplane Monthly March 1975, January 1996, March and June 1997, March and April 1999
Air Pictorial April 1996
Planes, Autumn 1982

INDEX

Note: This index does not include the Appendices.